VAISHNAVA PHILOSOPHY

Editor-in-Chief

Swami Chidatman Jee Maharaj

ANMOL PUBLICATIONS PVT. LTD.

NEW DELHI - 110 002 (INDIA)

ANMOL PUBLICATIONS PVT. LTD.
H.O.: 4374/4B, Ansari Road, Daryaganj,
New Delhi-110 002 (India)
Ph.: 23278000, 23261597
B.O.: No. 1015, Ist Main Road, BSK IIIrd Stage
IIIrd Phase, IIIrd Block,
Bangalore - 560 085 (India)
Visit us at: www.anmolpublications.com

Vaishnava Philosophy

First Published, 2009
ISBN 978-81-261-3624-7

PRINTED IN INDIA

Printed at Mehra Offset Press, Delhi.

Contents

Preface

India colourful and vibrant, a land as diverse as its people. A mosaic of faiths, cultures, customs and languages that blend harmoniously to form a composite whole. One of the world's oldest living civilizations—which gave to the world—the concept of zero, the primordial sound Aum..., Yoga and Buddhism.

She is old. She is young. The two diametrically opposite, different yet unique and dynamic faces of India coexist even at this turn of the millennium. With a history of heritage, learning and civilisation dating back to almost 5000 years or perhaps even more, the modern India presents itself as an ideal country. From snow clad mountains to breezy sparkling blue sea side. From arid deserts with shifting sand dunes to lush green country sides. From ancient relics and architectural ruins dating back into antiquity to modern cities, India has it all.

To the north it is bordered by the world's highest mountain chain, where foothill valleys cover the northernmost of the country's maximum states. Further south, plateaus, tropical rain forests and sandy deserts are bordered by palm fringed beaches. Its southernmost tip is the meeting point of three Oceans.

India is a land of staggering contrast with a mingling of the tradition and modernity that is an unique experience to be in India. One will forever remember the stay in this wonderful Homeland-a land truly of mystery and enchantment. Only India can offer such an astonishing variety of contrasts.

She has given birth to numerous spiritual leaders and founders of some of the popular religions of the present day world. India probably has the most religious diversity in any country. It's the birthplace of Hinduism, Buddhism, Sikhism and Jainism and is among the few places to have a resident

Zoroastrian population. India with her veritable treasure trove of culture, mysticism, philosophy, art, music and architecture to name a few. Much of India's classical music is devotional and the North Indian Hindustani and South Indian Carnatic streams are distinct and both have a complex 'raga' framework. The legacy of dance in India is tremendous. The classical dances of India are numerous such as Kathakali of Kerala, Bharatnatyam of Tamil Nadu, Kuchipudi of Andhra Pradesh, Manipuri and Odissi from Orissa are the prominent dance forms in this country that sways to an altogether novel beat. The Yakshagana, nautanki and puppetry are ancient folk forms that live on till date. The earliest specimens of Indian painting are the ones on the walls of the Ajanta Caves dating back to 2nd century BC. The Mughals had a huge impact on Indian art. The influence of Persian art brought placid garden scenes, illustrations from myths, legends and history into Indian art. Word craft, handicrafts, architecture and sculpture all contribute to this rich and varied domain. Indian literature, both in English and in the vernacular, is ever more popular around the world.

The book aims at presenting perspectives of different aspects of Indian culture. It has topics on Indian classical music and dance, painting, ayurveda and yoga which are traditional and indigenous system of physical and mental wellbeing, language, philosophy and excellence achievements by Indian in the different branches of science and social sciences. It also includes our ancient religious and classic literature.

It is hoped that this book will be welcomed by the general readers as also scholars who are interested in the Indian heritage.

—Editors

1

Vishnu and Vaishnavism

Vishnu: God of Preservation

Vishnu represents that aspect of the divine reality that maintains and sustains the creation. The endless creative force of Brahma gives life to all beings, and Vishnu's infinite sustaining force protects this life. Just as the feminine Supreme Energy, Vishnu has a "thousands names" which symbolize in fact just as many of his aspects. The daily repetition of Vishnu's thousand names is a living proof of his devotee's faith and ardour.

Vishnu's blue body is often depicted in blue. The colour blue symbolizes the infinite. Vishnu is represented as an infinite force. He has no form, no name, and he is incommensurable.

The garland of flowers around Vishnu's neck is a symbol of God's worship. The precious stone decorating his neck indicates that Vishnu fulfils his worshipper's desires and the crown is a symbol of the Divine's power and supreme authority. His two earrings stand for the dual nature of creation.

Vishnu wears yellow clothing. The yellow colour is associated with the terrestrial life. The dress of Vishnu, therefore, symbolizes his incarnations as a man, fighting for justice and destruction of the evil.

"Vishnu, the Preserver, sustains the whole creation and has the power of manifesting himself under numerous forms. In the great Cosmic Ocean, He sustains both the sun of the infinite and the eternal spirit of existence, which is the master of the universe."

Vishnu Purana

Vishnu the Preserver, is often depicted as lying on Ananta, a many-headed serpent. He lives on Meru with his wife Lakshmi. Vishnu usually rides through the heavens on Garuda, a man-bird, endowed with highly unusual powers, which spreads courage and Vedic knowledge.

Very often, Vishnu appears under the form of a young, very handsome man, with four arms (several arms is an indication of the divine, showing that they can perform several functions in the same time).

In his hands, Vishnu is shown holding a shell, a disk, a lotus, and a sceptre. The symbolism of the shell is two-folded: on one hand it represents Vishnu's relationship with the primordial waters and its use as a musical instrument, reminiscent of the primordial waters, and on the other hand the shell represents the five elements, and consequently the origin of manifestation.

The lotus is a symbol of the spiritual development and cosmic harmony. It also signifies the appearance of life on the neutral immensity of the primordial waters.

The opening of the lotus bud represents the realization of the possibilities contained in the "seed" of the human being. The sceptre is a symbol of power and authority.

The disk is a solar symbol, and the classical image of Vishnu, the ascendant, unifying, intellectual tendency of a human being. It represents the power that destroys the ignorance and the darkness, just as it is a symbol of a killing weapon, and of the sun. It destroys the evil through the process of illumination.

Vishnu is more a god of love. For a Vaishnavite, salvation is usually won by bhakti, a loving devotion to God as preserver. Vishnu is considered the guardian of mankind and the keeper of the Dharma. He preserves the order in the universe. Every time the humanity decays, Vishnu takes on the form of an avatar, a divine model coming on earth, to show people new paths for spiritual development. He has taken nine incarnations.

His first incarnation was a fish, Matsya; the second a turtle, Kurma; the third a wild boar, Varaha; the fourth a lion; the fifth a dwarf, Vamana; the sixth a warrior, Parasurama; the seventh a hero, Rama; the eighth Krishna; the ninth Buddha.

Vishnu's tenth incarnation, Kalki, is supposed to come about at the end of this age, Kali Yuga. He is said to be the cosmic rider of Apocalypse, who will take on the role of destroyer of what is evil on earth.

The most important incarnations of Vishnu are Rama and Krishna.

To Hindus, Rama, who is nearly always noble and righteous, is the ideal man. Sita, incarnation of Lakshmi and his loyal wife, is the ideal woman. All sects reverence Rama. No name is more commonly given to Hindu children, and all Hindus would like to die with his name on their lips. When Mahatma Gandhi fell, mortally wounded by a fanatic assassin, he murmured "Ai Ram, Ai Ram" (O Rama, O Rama).

Krishna, even more popular and lovable, is the subject of countless stories and legends. As a youth Krishna flirts with the Gopi milk maidens and has an affair with one, Radha. These erotic experiences are interpreted symbolically. Individual souls are drawn to God as the milkmaidens were attracted to Krishna, and one should give one's self to God in complete surrender as Radha did to Krishna.

Vaishnavism in the South

The ancient Vaishnava mystics and saints in the South were known as Alvars, and the Vaishnavism teachers as Acharayas. They had a powerful exponent of these views in Ramanuja, who attacked the Advaita interpretation of the Upanishads and gave recognition to three ultimate realities, God, Soul and Matter, the last two being dependent on the first.

As early as the 2nd century B.C. the renowned Besnagar Column had been erected by a Greek named Heliodorous, who had been converted to the Bhagavata or Vaishnava faith of which the Pancharatra doctrines then formed an integral part;

its scriptures were Satvata Samhita, the Mahabharata, and the Bhagavata and Vishnu Puranas. The origin of the Pancharatra doctrines which form the basis of Sri Vaishnava culture has be Pancharatra entraced further back to the well known Purusasukta of the Rig-Veda. The Satapatha Brahmana refers to the Pancharatra sacrifices performed by the primeval Narayana, the idea of Nara and Narayana (Primordial man and the deity Vishnu) being an integral part of ancient Indian thought. There are more than a dozen Vaishnava Upanishads. It was in the period from the 10th century up to the 17th that many Vaishnava works were produced. The Vaishnavas regard the Pancharatra literature as almost equal to the Vedas.

The Vaishnava Samhitas and other works insist on knowledge of and devotion to, the supreme Godhead rather than on Vedic studies or sacrifices. It is worthy of note that in the Bhagavata Purana (11th Skanda) the Alwars were prefigured or adverted to; several great devotees of Vishnu, the Purana states, would appear on the banks of the Tamraparni, Kritamala (Vaigai), Payasvin (Palar), Kaveri (Cauvery), and Mahanadi (Periyar).

The Alvars lived between the 5th and the 12th centuries. The first group included Saroyogin or Poygaiyal var, Bhatayogin or Bhutattalvar, Mahadyogin or Peyalvr and Bhaktisara or Tirumalisai-Puran. Nammalvar or Satakopa, who came in the next group, was perhaps the greatest of the Alvars. Others in this group included Madhurakaviyalvar, Kulasekhara Perumal, Vishnuchitta (or Periyalvar) and Andal, his adopted daughter. In the last of the groups were Bhaktanghrirenu (Tondaradippodiyalvar), Yogivahana (Tiruppanalvar) and Parakala (Tirumangaiyalvar). The Divya Prabhandha constitutes the collection of the Alvars' compositions in the Tamil language.

Vaishnavism in the North

One of the most influential Vaishnava cults was founded by Vallabhacarya, a Telugu Brahmin who lived in the 15th century. He migrated to the North and in his numerous works in the North he gave an interpretation of the Vedanta differing from that of Ramanuja, as also of Shankara. He called his doctrine Suddha Advaita, pure non-dualism. The world is real, and not an illusion God is Nimitta-Karana, the causative being.

Discarding the Maya theory Vallabhacharya asserts that God cannot be described by negatives but only by his holy and gracious attributes, and is personified in Krishna He is not only karta, creator, but also Bhokta, enjoyer. Though he has no need to assume a bodily form, he often does so to please his devotees.

Regarding Bhakti as the chief means of salvation and superior to Jnana, (knowledge) Vallabha opposed all kinds of asceticism. The body is the temple of God, he said. The famous Upanisadic precept Tatvamasi was by an ingenious interpretation, modified by Vallabha as Atatvamasi, "That thou art not". Vallabhacharya's doctrines were fully interpreted and expounded by his son Vitthala.

Later, in Northern India, there arose the Chaitanya movement. Nimbarka had already elevated Radha, the consort of Krishna, to the highest position. Jayadeva, the author of Gita-Govinda, and other poets like Vidyapati, Umapati and Chandidas, adopted the Radha-Krishna cult.

Chaitanya, the great Vaishnava teacher of the 15th century transformed the Vaishnava faith and extended his influence in most parts of Northern India. He accepted converts from Islam, the foremost among them being Haridas, Rupa and Sanatana. Salvation, according to his doctrine, consists in the eternal experience of God's love. Chaitanya exercised great influence over later Indian thought.

Chaitanya Mahaprabhu was born in Mayapur in the town of Nadia just after sunset on the evening of the 23rd Phalguna 1407 Sakabda, answering to the 18th of February, 1486, of the Christian Era. The moon was eclipsed at the time of his birth, and the people of Nadia were then engaged, as was usual on such occasions, in bathing in the Bhagirathi with loud cheers of Haribol. His father, Jagannath Mishra, a poor brahmana of the Vedic order, and his mother, Sachi-devi, a model good woman, both descended from brahmana stock originally residing in Sylhet. Mahaprabhu was a beautiful child, and the ladies of the town came to see him with presents. His mother's father,

Pandita Nilambara Chakravarti, a renowned astrologer, foretold that the child would be a great personage in time; and he, therefore, gave him the name Vishvambhara. The ladies of the neighbourhood styled him Gaurahari on account of his golden complexion, and his mother called him Nimai on account of the nimba tree near which he was born. Beautiful as the lad was, everyone heartily loved to see him every day. As he grew up he became a whimsical and frolicsome lad. After his fifth year, he was admitted into a pathasala where he picked up Bengali in a very short time.

Most of his contemporary biographers have mentioned certain anecdotes regarding Chaitanya which are simple records of his early miracles. It is said that when he was an infant in his mother's arms he wept continually, and when the neighbouring ladies cried Haribol he used to stop. Thus there was a continuation of the utterance of Haribol in the house, foreshewing the future mission of the hero. It has also been stated that when his mother once gave him sweetmeats to eat, he ate clay instead of the food. His mother asking for the reason, he stated that as every sweetmeat was nothing but clay transformed, he could eat clay as well. His mother, who was also the consort of a Pandita, explained that every article in a special state was adapted to a special use.

Earth, while in the state of a jug, could be used as a water pot, but in the state of a brick such a use was not possible. Clay, therefore, in the form of sweetmeats was usable as food, but clay in its other states was not. The lad was convinced and admitted his stupidity in eating clay and agreed to avoid the mistake in the future. Another miraculous act has been related. It is said that a brahmana on pilgrimage became a guest in his house, cooked food and read grace with meditation upon Krishna. In the meantime the lad came and ate up the cooked rice. The brahmana, astonished at the lad's act, cooked again at the request of Jagannath Mishra. The lad again ate up the cooked rice while the brahmana was offering the rice to Krishna with meditation. The brahmana was persuaded to cook for the third time. This time all the inmates of the house had fallen asleep, and the lad shewed himself as Krishna to the traveller and blessed him. The brahmana was then lost in ecstasy at the appearance of the object of his worship. It has also been stated that two thieves stole away the lad from his father's door with a view to purloin his jewels and gave him sweetmeats on the way.

The lad exercised his illusory energy and deceived the thieves back towards his own house. The thieves, for fear of detection, left the boy there and fled. Another miraculous act that has been described is the lad's demanding and getting from Hiranya and Jagadisha all the offerings they had collected for worshipping Krishna on the day of Ekadasi. When only four years of age he sat on rejected cooking pots which were considered unholy by his mother. He explained to his mother that there was no question of holiness and unholiness as regards earthen pots thrown away after the cooking was over. These anecdotes relate to his tender age up to the fifth year.

In his eighth year, he was admitted into the tola of Gangadasa Pandita in Ganganagara close by the village of Mayapur. In two years he became well read in Sanskrit grammar and rhetoric. His readings after that were of the nature of self-study in his own house, where he had found all-important books belonging to his father, who was a Pandita himself. It appears that he read the smriti in his own study, and the nyaya also, in competition with his friends, who were then studying under the celebrated Pandita Raghunatha Shiromani.

Now, after the tenth year of his age, Chaitanya became a passable scholar in grammar, rhetoric, the smriti and the nyaya. It was after this that his elder brother Visvarupa left his house and accepted the ashrama (status) of a sanyasi (ascetic). Chaitanya, though a very young boy, consoled his parents, saying that he would serve them with a view to please God. Just after that, his father left this world. His mother was exceedingly sorry, and Mahaprabhu, with his usual contented appearance, consoled his widowed mother.

It was at the age of 14 or 15 that Mahaprabhu was married to Lakshmidevi, the daughter of Vallabhacharya, also of Nadia. He was at this age considered one of the best scholars of Nadia, the renowned seat of nyaya philosophy and Sanskrit learning.

Not to speak of the smarta panditas, the Naiyayikas were all afraid of confronting him in literary discussions. Being a married man, he went to Eastern Bengal on the banks of the Padma for acquirement of wealth. There he displayed his learning and obtained a good sum of money.

It was at this time that he preached Vaishnavism at intervals. After teaching him the principles of Vaishnavism, he ordered Tapana Mishra to go to and live in Benares. During his residence in East Bengal, his wife Lakshmidevi left this world from the effects of snakebite. On returning home, he found his mother in a mourning state. He consoled her with a lecture on the uncertainty of human affairs. It was at his mother's request that he married Vishnupriya, the daughter of Raja Pandita Sanatana Mishra. His comrades joined him on his return from pravasa or sojourn. He was now so renowned that he was considered to be the best Pandita in Nadia. Keshava Mishra of Kashmir, who had called himself the Great Digvijayi, came to Nadia with a view to discuss with the panditas of that place. Afraid of the so-called conquering Pandita, the tola professors of Nadia left their town on pretence of Invitation. Keshava met Mahaprabhu at the Barokona-ghata in Mayapur, and after a very short discussion with him he was defeated by the boy, and mortification obliged him to decamp. Nimai Pandita was now the most important Pandita of his times.

It was at the age of 16 or 17 that he travelled to Gaya with a host to sing the holy name of Hari in the streets and bazaars. This created a sensation and roused different feelings in different quarters. The bhaktas were highly pleased. The smarta brahmanas became jealous of Nimai Pandita's success and complained to Chand Kazi against the character of Chaitanya as un-Hindu. The Kazi came to Srivasa Pandita's house and broke a mridanga (khola drum) there and declared that unless Nimai Pandita ceased to make noise about his queer religion he would be obliged to enforce Mohammedanism on him and his followers. This was brought to Mahaprabhu's notice. He ordered the townspeople to appear in the evening, each with a torch in his hand. This groups, and on his arrival in the Kazi's house, he held a long conversation with the Kazi and in the end communicated into his heart his Vaishnava influence by touching his body. The Kazi then wept and admitted that he had felt a keen spiritual influence which had cleared up his doubts and produced in him a religious sentiment which gave him the highest ecstasy. The Kazi then joined the sankirtana party. The world was astonished at the spiritual power of the Great Lord, and hundreds and hundreds of heretics converted and joined the banner of Vishvambhara after this affair.

It was after this that some of the jealous and low-minded brahmanas of Kulia picked a quarrel with Mahaprabhu and collected a party to oppose him. Nimai Pandita was naturally a soft-hearted person, though strong in his principles. He declared that party feelings and sectarianism were the two great enemies of progress and that as long as he should continue to be an inhabitant of Nadia belonging to a certain family, his mission would not meet with complete success. He then resolved to be a citizen of the world by cutting his connection with his particular family, caste and creed, and with this resolution he embraced the position of a sanyasi at Katwa, under the guidance of Keshava Bharati of that town, on the 24th year of his age. His mother and wife wept bitterly for his separation, but our hero, though soft in heart, was a strong person in principle. He left his little world in his house for the unlimited spiritual world of Krishna with man in general.

After his sannyasa, he was induced to visit the house of Advaita Prabhu in Santipura. Advaita managed to invite all his friends and admirers from Nadia and brought Sachidevi to see her son. Both pleasure and pain invaded her heart when she saw her son in the attire of a sanyasi. As a sanyasi, Krishna Chaitanya put on nothing but a kaupina and a bahirvasa (outer covering). His head was without hair, and his hands bore a danda (stick) and a kamandalu (hermit's water pot). The holy son fell at the feet of his beloved mother and said, "Mother! This body is yours, and I must obey your orders. Permit me to go to Vrindavana for my spiritual attainments." The mother, in consultation with Advaita and others, asked her son to reside in Puri (the town of Jagannath) so that she might obtain his information now and then.

Mahaprabhu agreed to that proposition and in a few days left Santipura for Orissa. His biographers have described the journey of Krishna Chaitanya (that was the name he got after his sannyasa) from Santipura to Puri in great detail. He travelled along the side of the Bhagirathi as far as Chatrabhoga, situated now in Thana Mathurapura, Diamond Harbour, 24 Parganas. There he took a boat and went as far as Prayaga-ghata in the Midnapura District. Thence he walked through Balasore and Cuttack to Puri, seeing the temple of Bhubaneshvara on his way. Upon his arrival at Puri he saw Jagannath in the temple and resided with Sarvabhauma at the request of the latter.

Sarvabhauma was a gigantic Pandita of the day. His readings knew no bounds. He was the best naiyayika of the times and was known as the most erudite scholar in the Vedanta philosophy of the school of Shankaracharya. He was born in Nadia (Vidyanagara) and taught innumerable pupils in the nyaya philosophy in his tola there. He had left for Puri some time before the birth of Nimai Pandita. His brother-in-law Gopinatha Mishra introduced our new sanyasi to Sarvabhauma, who was astonished at his personal beauty and feared that it would be difficult for the young man to maintain sannyasa-dharma during the long run of his life. Gopinatha, who had known Mahaprabhu from Nadia, had a great reverence for him and declared that the sanyasi was not a common human being.

On this point Gopinatha and Sarvabhauma had a hot discussion. Sarvabhauma then requested Mahaprabhu to hear his recitation of the Vedanta-sutras, and the latter tacitly submitted. Chaitanya heard with silence what the great Sarvabhauma uttered with gravity for seven days, at the end of which the latter said, "Krishna Chaitanya! I think you do not understand the Vedanta, for you do not say anything after hearing my recitation and explanations." The reply of Chaitanya was that he understood the sutras very well, but he could not make out what Shankaracharya meant by his commentaries. Astonished at this, Sarvabhauma said, "How is it that you understand the meanings of the sutras and do not understand the commentaries which explain the sutras? All well! If you understand the sutras, please let me have your interpretations."

Mahaprabhu thereon explained all the sutras in his own way without touching the pantheistic commentary of Shankara.

The keen understanding of Sarvabhauma saw the truth, beauty and harmony of arguments in the explanations given by Chaitanya and obliged him to utter that it was the first time that he had found one who could explain the Brahma-sutras in such a simple manner. He admitted also that the commentaries of Shankara never gave such natural explanations of the Vedanta-sutras as he had obtained from Mahaprabhu. He then submitted himself as an advocate and follower. In a few days Sarvabhauma turned out to be one of the best Vaishnavas of the time. When reports of this came out, the whole of Orissa sang the praise of Krishna Chaitanya, and hundreds and hundreds came to him and became his followers. In the meantime Mahaprabhu thought of visiting Southern India, and he started with one Krishnadasa Brahmana for the journey.

His biographers have given us a detail of the journey. He went first to Kurmaksetra, where he performed a miracle by curing a leper named Vasudeva. He met Ramananda Raya, the Governor of Vidyanagara, on the banks of the Godavari and had a philosophical conversation with him on the subject of prema-bhakti. He worked another miracle by touching (making them immediately disappear) the seven tala trees through which Ramachandra, the son of Dasaratha, had shot his arrow and killed the great Bali Raja. He preached Vaishnavism and nama-sankirtana throughout the journey.

At Rangaksetra he stayed for four months in the house of one Venkata Bhatta in order to spend the rainy season. There he converted the whole family of Venkata from Ramanuja Vaishnavism to Krishna-bhakti, along with the son of Venkata, a boy of ten years named Gopala, who afterwards came to Vrndavana and became one of the six Gosvamis or prophets serving under their leader Sri Krishna Chaitanya. Trained up in Sanskrit by his uncle Prabodhananda Saraswati, Gopala wrote several books on Vaishnavism.

Chaitanya visited numerous places in Southern India as far as Cape Comorin and returned to Puri in two years by

Pandepura on the Bhima. In this latter place he spiritualized one Tukarama, who became from that time a religious preacher himself. This fact has been admitted in his abhangas, which have been collected in a volume by Mr. Satyendra Nath Tagore of the Bombay Civil Service. During his journey he had discussions with the Buddhists, the Jains and the Mayavadis in several places and converted his opponents to Vaishnavism.

Upon his return to Puri, Raja Prataparudra-deva and several Pandita brahmanas joined the banner of Chaitanya Mahaprabhu. He was now twenty-seven years of age. In his twenty-eighth year he went to Bengal as far as Gauda in Malda. There he picked up two great personages named Rupa and Sanatana. Though descended from the lines of the Karnatic brahmanas, these two brothers turned demi-Moslems by their continual contact with Hussain Shah, the then Emperor of Gauda. Their names had been changed by the Emperor into Dabira Khasa and Sakara Mallika, and their master loved them heartily since they were both learned in Persian, Arabic and Sanskrit and were loyal servants of the state. The two gentlemen had found no way to come back as regular Hindus and had written to Mahaprabhu for spiritual help while he was at Puri. Mahaprabhu had written in reply that he would come to them and extricate them from their spiritual difficulties. Now that he had come to Gauda, both the brothers appeared before him with their long-standing prayer. Mahaprabhu ordered them to go to Vrndavana and meet him there.

Chaitanya returned to Puri through Santipura, where he again met his dear mother. After a short stay at Puri he left for Vrndavana. This time he was accompanied by one Balabhadra Bhattacarya. He visited Vrndavana and came down to Prayaga (Allahabad), converting a large number of Mohammedans to Vaishnavism by argument from the Koran. The descendants of those converts are still known as Pathana Vaishnavas. Rupa Gosvami met him at Allahabad. Chaitanya trained him up in spirituality in ten days and directed him to go to Vrndavana on missions. His first mission was to write theological works explaining scientifically pure bhakti and prema. The second mission was to revive the places where

Krishnacandra had in the end of Dvapara-yuga exhibited His spiritual lila (pastimes) for the benefit of the religious world. Rupa Gosvami left Allahabad for Vrndavana, and Mahaprabhu came down to Benares.

There he resided in the house of Chandrashekhar and accepted his daily bhiksa (meal) in the house of Tapana Mishra. Here it was that Sanatana Gosvami joined him and took instruction for two months in spiritual matters. The biographers, especially Krishnadasa Kaviraja, have given us details of Chaitanya's teachings to Rupa and Sanatana. Krishnadasa was not a contemporary writer, but he gathered his information from the Gosvamis themselves, the direct disciples of Mahaprabhu. Jiva Gosvami, who was nephew of Sanatana and Rupa and who has left us his invaluable work the Sat-sandarbha, has philosophized on the precepts of his great leader. We have gathered and summarized the precepts of Chaitanya from the books of those great writers.

While at Benares, Chaitanya had an interview with the learned sannyasis of that town in the house of a Maratha brahmana who had invited all the sannyasis for entertainment. At this interview, Chaitanya shewed a miracle which attracted all the sannyasis to him. Then ensued reciprocal conversation. The sannyasis were headed by their most learned leader Prakasananda Saraswati. After a short controversy, they submitted to Mahaprabhu and admitted that they had been misled by the commentaries of Shankaracharya.

It was impossible even for learned scholars to oppose Chaitanya for a long time, for there was some spell in him which touched their hearts and made them weep for their spiritual improvement. The sannyasis of Benares soon fell at the feet of Chaitanya and asked for his grace (krpa). Chaitanya then preached pure bhakti and instilled into their hearts spiritual love for Krishna which obliged them to give up sectarian feelings. The whole population of Benares, on this wonderful conversion of the sannyasis, turned Vaishnavas, and they made a master sankirtana with their new Lord. After sending Sanatana to Vrndavana, Mahaprabhu went to Puri again through the jungles with his comrade Balabhadra. Balabhadra reported that Mahaprabhu had shown a good many miracles on his way to Puri, such as making tigers and elephants dance on hearing the name of Krishna.

From this time, that is, from his 31st year, Mahaprabhu continually lived in Puri in the house of Kasi Mishra until his disappearance in his forty-eighth year at the time of sankirtana in the temple of Tota-gopinatha. During these 18 years, his life was one of settled love and piety. He was surrounded by numerous followers, all of whom were of the highest order of Vaishnavas and who were distinguished from the common people by their purest character and learning, firm religious principles and spiritual love of Radha-Krishna. Svarupa Damodara, who had been known by the name of Purushottamacarya while Mahaprabhu was in Nadia, joined him from Benares and accepted service as his secretary. No production of any poet or philosopher could be laid before Mahaprabhu unless Svarupa had passed it as pure and useful. Raya Ramananda was his second mate.

Both he and Svarupa would sing while Mahaprabhu expressed his sentiments on a certain point of worship. Paramananda Puri was his minister in matters of religion. There are hundreds of anecdotes described by his biographers which we do not think it meet here to reproduce. Mahaprabhu slept short. His sentiments carried him far and wide in the firmament of spirituality every day and night, and all his admirers and followers watched him throughout. He worshipped, communicated with his missionaries at Vrndavana, and conversed with those religious men who newly came to visit him. He sang and danced, took no care of himself and oft-times lost himself in religious beatitude.

All who came to him believed in him as the all-beautiful God appearing in the nether world for the benefit of mankind. He loved his mother all along and sent her mahaprasada now and then with those who went to Nadia. He was most amiable in nature. Humility was personified in him. His sweet appearance gave cheer to all who came in contact with him. He appointed Prabhu Nityananda as the missionary in charge of Bengal. He dispatched six disciples (Gosvamis) to Vrndavana to preach love in the upcountry. He punished all of his disciples who deviated from a holy life. This he markedly did in the case of Junior Haridasa. He never lacked in giving proper instructions in life to those who solicited them. This will be seen in his teachings to Raghunatha dasa Gosvami. His treatment to Haridasa (senior) will show how he loved spiritual men and how he defied caste distinction in spiritual brotherhood.

Lord Chaitanya's Mission

Lord Chaitanya Mahaprabhu instructed His disciples to write books on the Science of Krishna, a task which those who follow Him have continued to carry out down to the present day. The elaborations and expositions on the philosophy taught by Lord Chaitanya are in fact most voluminous, exacting and consistent due to the system of disciplic succession. Although Lord Chaitanya was widely renowned as a scholar in His youth, He left only eight verses, called Siksastaka. These eight verses clearly reveal His mission and precepts. These supremely valuable prayers are translated herein.

1. Glory to the Sri Krishna sankirtana, which cleanses the heart of all the dust accumulated for years and extinguishes the fire of conditional life, of repeated birth and death. This sankirtana movement is the prime benediction for humanity at large because it spreads the rays of the benediction moon. It is the life of all transcendental knowledge. It increases the ocean of transcendental bliss, and it enables us to fully taste the nectar for which we are always anxious.
2. O my Lord, Your holy name alone can render all benediction to living beings, and thus You have hundreds and millions of names like Krishna and Govinda. In these transcendental names You have invested all Your transcendental energies. There are not even hard and fast rules for chanting these names. O my Lord, out of kindness You enable us to easily approach You by Your holy names, but I am so unfortunate that I have no attraction for them.

3. One should chant the holy name of the Lord in a humble state of mind, thinking oneself lower than the straw in the street; one should be more tolerant than a tree, devoid of all sense of false prestige and should be ready to offer all respect to others. In such a state of mind one can chant the holy name of the Lord constantly.
4. O almighty Lord, I have no desire to accumulate wealth, nor do I desire beautiful women, nor do I want any number of followers. I only want Your causeless devotional service birth after birth.
5. O son of Maharaja Nanda [Krishna], I am Your eternal servitor, yet somehow or other I have fallen into the ocean of birth and death. Please pick me up from this ocean of death and place me as one of the atoms at Your lotus feet.
6. O my Lord, when will my eyes be decorated with tears of love flowing constantly when I chant Your holy name? When will my voice choke up, and when will the hairs of my body stand on end at the recitation of Your name?
7. O Govinda! Feeling Your separation, I am considering a moment to be like twelve years or more. Tears are flowing from my eyes like torrents of rain, and I am feeling all vacant in the world in Your absence.
8. I know no one but Krishna as my Lord, and He shall remain so even if He handles me roughly by His embrace or makes me brokenhearted by not being present before me. He is completely free to do anything and everything, for He is always my worshipful Lord unconditionally.

[Originally delivered as five morning lectures on Chaitanya-caritamrta—the authoritative biography of Lord Chaitanya Mahaprabhu, by Krishnadasa Kaviraja Gosvami—before the International Society for Krishna Consciousness, New York City, April 10-14, 1967.]

The word Chaitanya means living force. As living entities, we can move, but a table cannot because it does not possess living force. Movement and activity may be considered to be signs or symptoms of the living force. Indeed, it may be said that there can be no activity without the living force. Although the living force is present in the material condition, it is not amrta, immortal. The words Chaitanya-caritamrta, then, may be translated as "the character of the living force in immortality."

But how is this living force displayed immortally? It is not displayed by man or any other creature in this material universe, for none of us are immortal in these bodies. We possess the living force, we perform activities, and we are immortal by our nature and constitution, but the material condition into which we have been put does not allow our immortality to be displayed.

It is stated in the Katha Upanisad that eternality and the living force belong both to ourselves and God. Although this is true in that both God and ourselves are immortal, there is a difference. As living entities, we perform many activities, but we have a tendency to fall down into material nature. God has no such tendency. Being all-powerful, He never comes under the control of material nature. Indeed, material nature is but one display of His inconceivable energies.

On the ground we may see only clouds in the sky, but if we fly above the clouds we can see the sun shining. From the sky, skyscrapers and cities seem very tiny; similarly, from God's position this entire material creation is insignificant. The tendency of the conditioned living entity is to come down from the heights where everything can be seen in perspective. God, however, does not have this tendency. The Supreme Lord is not subject to fall down into illusion (maya) any more than the sun is subject to fall beneath the clouds. Because the Supreme Lord is not subject to illusion, He is unconditioned; because we, as finite living entities, are prone to fall into illusion, we are called conditioned.

Impersonalist philosophers (Mayavadis) maintain that both the living entity and God Himself are under the control of maya when they come into this material world. This may be true of the living entity, but it is not true of God, for in all instances the material energy is working under His direction. Those who consider the Supreme Lord to be subject to material conditioning are called fools by Krishna Himself in Bhagwad Gita:

avajananti mam mudha
manusim tanum asritam
param bhavam ajananto
mama bhuta-mahesvaram

"Fools deride Me when I descend in the human form. They do not know My transcendental nature and My supreme dominion over all that be." (Bg. 9.11)

Lord Chaitanya Mahaprabhu should not be considered to be one of us. He is Krishna Himself, the supreme living entity, and as such He never comes under the cloud of maya. Krishna, His expansions, and even His higher devotees never fall into the clutches of illusion. Lord Chaitanya came to earth simply to preach Krishna-bhakti, love of Krishna. In other words, He is Lord Krishna Himself teaching the living entities the proper way to approach Krishna. He is like a teacher who, seeing a student doing poorly, takes up a pencil and writes, saying, "Do it like this: A, B, C." By this, one must not foolishly think that the teacher is learning his ABC's. Although He appears in the guise of a devotee, we should always remember that Lord Chaitanya is Krishna (God) Himself teaching us how to become Krishna conscious, and we must study Him in that light.

In Bhagwad Gita Lord Krishna sets forth the highest religious principle in this way:

sarva-dharman parityajya
mam ekam saranam vraja
aham tvam sarva-papebhyo
mokshayisyami ma sucah

"Abandon all varieties of religion and just surrender unto Me. I shall deliver you from all sinful reaction. Do not fear." (Bg. 18.66)

This may seem to be a simple instruction to follow, but invariably our reaction is, "Oh, surrender? Give up? But I have so many responsibilities." And maya, illusion, says to us, "Don't do it, or you'll be out of my clutches. Just stay in my clutches, and I'll kick you." It is a fact that we are constantly being kicked by maya, just as the male ass is kicked in the face by the she-ass when he comes for sex. Similarly, cats and dogs are always fighting and whining when they have sex. These are the tricks of nature. Even an elephant in the jungle is caught by the use of a trained she-elephant who leads him into a pit. Maya has many activities, and in the material world her strongest shackle is the female.

Of course in actuality we are neither male nor female—for these designations refer only to the outer dress, the body. We are all actually Krishna's servants. In conditioned life, however, we are shackled by the iron chains which take the form of beautiful women. Thus every male is bound by sex life, and therefore when one attempts to gain liberation from the material clutches, he must first learn to control the sex urge. Unrestricted sex puts one fully in the clutches of illusion. Lord Chaitanya Mahaprabhu officially renounced this illusion at the age of twenty-four, although His wife was sixteen and His mother seventy, and He was the only male member of the family. Although He was a brahmana and was not rich, He took sannyasa, the renounced order of life, and thus extricated Himself from family entanglement.

If we wish to become fully Krishna conscious, we have to give up the shackles of maya, or, if we remain with maya, we should live in such a way that we will not be subject to illusion. It is not necessary for one to abandon his family, for there were many householders amongst Lord Chaitanya's closest devotees. What must be renounced is the propensity for material enjoyment. Although Lord Chaitanya approved of a householder having regulated sex in marriage, He was very strict with those in the renounced order, and He even banished junior Haridasa for glancing lustfully at a young woman.

The point is that one must take up a particular path and stick to it, obeying all the rules and regulations necessary for success in spiritual life. It was Lord Chaitanya's mission that He teach the path of Krishna consciousness to all men and thereby enable them to partake of the immortality of spiritual life.

From Chaitanya-caritamrta we learn how Chaitanya taught people to become immortal, and thus the title may be properly translated as "the immortal character of the living force." The supreme living force is the Supreme Personality of Godhead. He is also the supreme entity. There are innumerable living entities, and all of them are individual. This is very easy to understand: We are all individual in thought and desires, and the Supreme Lord is also an individual person. He is different, though, in that He is the leader, the one whom no one can excel. Amongst the created living entities, one being can excel another in one capacity or another.

The Lord is an individual, just as the living entities are individual, but He is different in that He is the supreme individual. God is also infallible, and in Bhagwad Gita He is addressed as Achyuta, which means, "He who never falls down." This is indicated because in Bhagwad Gita Arjuna had fallen into delusion but Krishna had not. We often hear it said that God is infallible, and in Bhagwad Gita Krishna states:

nanyam gunebhyah kartaram
yada drastanupasyati
gunebhyas ca param vetti
mad-bhavam so 'dhigacchati

"When you see that there is nothing beyond these modes of nature in all activities and that the Supreme Lord is transcendental to all these modes, then you can know My spiritual nature." (Bg. 14.19)

Thus we should not think that Krishna is overpowered by the material potency when He is in the material world. Krishna and His incarnations are not under the control of material nature. They are totally free. Indeed, in Srimad-Bhagavatam one who has a godly nature is actually defined as one who is not affected by the modes of material nature, although in material nature. If even a devotee can attain this freedom, then what to speak of the Supreme?

The real question is how can we remain unpolluted by material contamination while in the material world. It was Rupa Gosvami who explained that we can remain uncontaminated while in the world if we simply make it our ambition to serve Krishna. One may then justifiably ask, "How can I serve?" Obviously this is not simply a matter of meditation, which is just an activity of the mind, but of practical work. Love of Krishna's service can only be attained by working for Krishna. In such work, we should leave no resource unused. Whatever is there, whatever we have, should be used for Krishna. We can use everything: typewriters, automobiles, airplanes, missiles—anything.

If we simply speak to people about Krishna consciousness, we are also rendering service. If our minds, senses, speech, money and energies are thus engaged in the service of Krishna, we cannot be considered to be existing in material nature. By virtue of spiritual consciousness, or Krishna consciousness, we transcend the platform of material nature. It is a fact that Krishna, His expansions and His devotees—that is, those who work for Him—are not in material nature, although people with a poor fund of knowledge think that they are.

Chaitanya-caritamrta teaches that the spirit soul is immortal and that our activities in the spiritual world are also immortal. The Mayavadis, who hold to the view that the Absolute is impersonal and formless, contend that a realized soul has no need to talk. However, the Vaishnavas, who are devotees of Krishna, contend that when one reaches the stage of realization, he really begins to talk. "Previously we only talked of nonsense," the Vaishnava says. "Now let us begin our real talks, talks of Krishna." The Mayavadis are also fond of using the example of the water pot, maintaining that when a pot is not filled with water it makes a sound, but that when it is filled it makes no sound. But are we waterpots? How can we be compared to them? A the material world. In this instance both the sun and moon have risen together, and it is proper to offer obeisances directly to Lord Chaitanya and Nityananda.

In the Western world where the glories of Lord Chaitanya are relatively unknown, one may inquire, "Who is Krishna Chaitanya?" The scriptural conclusion in answer to that question is that He is the Supreme Personality of God head. Generally in the Upanishads the Supreme Absolute Truth is described in an impersonal way, but the personal aspect of the Absolute Truth is mentioned in the Isopanisad, where, after a description of the all-pervading, we find the following verse:

hiranmayena patrena
satyasyapihitam mukham
tat tvam pusann apavrnu
satya-dharmaya drstaye

"O my Lord, sustainer of all that lives, Your real face is covered by Your dazzling effulgence. Kindly remove that covering and exhibit Yourself to Your pure devotee." (Sri Isopanisad, Mantra 15)

The impersonalists do not have the power to go beyond the effulgence of God and arrive at the personality from whom this effulgence is emanating. At the end of Isopanisad, however, there is a hymn to the Personality of Godhead. It is not that the impersonal Brahman is denied; it is also described, but that Brahman is considered to be the glaring effulgence of the body of Chaitanya. In other words, Krishna Chaitanya is the basis of the impersonal Brahman.

It is also stated by Krishna in Bhagwad Gita that the impersonal Brahman rests on Him (brahmano hi pratisthaham, Bg. 14.27). The Paramatma, or Supersoul, which is present within the heart of every living entity and within every atom of the universe, is but the partial representation of Chaitanya. Krishna Chaitanya is therefore the basis of Brahman and the Supreme Personality of Godhead as well. As the Supreme, He is full in six opulences: wealth, fame, strength, beauty, knowledge and renunciation. In short, we should know that He is Krishna, God, and nothing is equal to or greater than Him. There is no superior to be conceived. He is the Supreme Person.

It was Rupa Gosvami, a confidential devotee taught for more than ten days continuously by Lord Chaitanya, who wrote:

namo maha-vadanyaya
Krishna-prema-pradaya te
Krishnaya Krishna-Chaitanya-
namne gaura-tvise namah

"I offer my respectful obeisances unto the Supreme Lord Sri Krishna Chaitanya, who is more magnanimous than any other avatara, even Krishna Himself, because He is bestowing freely what no one else has ever given—pure love of Krishna."

It is not that Chaitanya teaches a long and elaborate path to God realization. He is completely spiritual, and He begins from the point of surrender to Krishna. He does not pursue the paths of karma-yoga or jnana-yoga or hatha-yoga but begins at the end of material existence, at the point where one gives up all material attachment. In Bhagwad Gita Krishna began His teachings by distinguishing the soul from matter and in the Eighteenth Chapter concluded at the point where the soul surrenders to Him in devotion.

The Mayavadis would have all talk cease there, but at that point the real discussion only begins. It is the Vedanta-sutra which begins: athato brahma-jijnasa: "Now let us begin to inquire about the Supreme Absolute Truth." Rupa Gosvami thus praises Chaitanya as the most munificent incarnation of all, for He gives the greatest gift by indicating the highest form of devotional service. In other words, He answers the most important inquiries that anyone can make.

There are different stages of devotional service and God realization. Strictly speaking, anyone who accepts the existence of God is situated in devotional service. To acknowledge that God is great is something, but not much. Chaitanya, preaching as an acarya, a great teacher, taught that we can enter into a relationship with God and actually become God's friend. In Bhagwad Gita Krishna showed Arjuna His universal form because Arjuna was His "very dear friend." Upon seeing Krishna as the Lord of the universes, however, Arjuna actually asked Krishna to forgive the familiarity of his friendship.

Chaitanya goes beyond this point. Through Lord Chaitanya we can become friends with Krishna, and there is no limit to this friendship. We can become friends of Krishna not in awe or adoration but in complete freedom. We can even relate to God as His father. This is not only the philosophy of Chaitanya-caritamrta but of Srimad-Bhagavatam as well. There are no other literatures in the world in which God is treated as the son of a devotee. Usually God is seen as the almighty Father who supplies the demands of His sons. The great devotees, however, sometimes treat God as a son in their execution of devotional service.

The son demands, and the father supplies, and in supplying Krishna the devotee becomes like a father. Instead of taking from God, we give to God. It was in this relationship that Krishna's mother, Yasoda, told the Lord, "Here, eat this or You'll die. Eat nicely." In this way Krishna, although the proprietor of everything, depends on the mercy of His devotee. This is a uniquely high level of friendship in which the devotee actually believes himself to be the father of Krishna.

However, Lord Chaitanya's greatest gift was His teaching that Krishna can be actually treated as one's lover. In this relationship the Lord is so much attached that He expresses His inability to reciprocate. Krishna was so obliged to the gopis, the cowherd girls of Vrndavana, that He felt unable to return their love. "I cannot repay your love," He told them. "I have no more assets to return." Thus devotional service is performed on this excellent platform, and knowledge of the devotee's relationship to Krishna as lover and beloved was given by Chaitanya Mahaprabhu. It was never given by any previous incarnation or acharya.

Thus Rupa Gosvami wrote of Chaitanya: "Devotional service itself is the highest platform, the glorious platform which You have contributed. You are Krishna in a yellow complexion, and You are Sacinandana, the son of mother Saci. Those who hear Chaitanya-caritamrta will keep You in their hearts. It will be easy to understand Krishna through You." Thus Chaitanya Mahaprabhu came to deliver Krishna. His method of deliverance was not meditation, fruitive activities or scriptural study, but love.

We have often heard the phrase "love of Godhead." How far this love of Godhead can actually be developed can be learned from the Vaishnava philosophy. Theoretical knowledge of love of God can be found in many places and in many scriptures, but what that love of Godhead actually is and how it is developed can be found in Vaishnava literatures. It is the unique and highest development of love of God that is given by Chaitanya Mahaprabhu.

Even in this material world we can have a little sense of love. How is this possible? It is due to the love which is found in the Godhead. Whatever we find within our experience within this conditional life is situated in the Supreme Lord, who is the ultimate source of everything. In our original relationship with the Supreme Lord there is real love, and that love is reflected pervertedly through material conditions. Our real love is continuous and unending, but because that love is reflected pervertedly in this material world, it lacks continuity and is inebriating. If we want real transcendental love, we have to transfer our love to the supreme lovable object—the Supreme Personality of Godhead. This is the basic principle of Krishna consciousness.

In material consciousness we are trying to love that which is not at all lovable. We give our love to dogs and cats, running the risk that at the time of death we may think of them and consequently take birth in a family of cats or dogs. Thus love that does not have Krishna as its object leads downward. It is not that Krishna or God is something obscure or something that only a few chosen people can attain. Chaitanya Mahaprabhu informs us that in every country and in every scripture there is some hint of love of Godhead. Unfortunately no one knows what love of Godhead actually is the Vedic scriptures, however, are different in that they can direct the individual in the proper way to love God. Other scriptures do not give information on how one can love God, nor do they actually define or describe what or who the Godhead actually is. Although they officially promote love of Godhead, they have no idea how to execute it. But Chaitanya Mahaprabhu gives a practical demonstration of how to love God in a conjugal relationship.

Taking the part of Radharani, Chaitanya tries to love Krishna as Radharani loved Him. Krishna was always amazed by Radharani's love. "How does Radharani give Me such pleasure?" He would ask. In order to study Radharani, Krishna lived in Her role and tried to understand Himself. This is the secret of Lord Chaitanya's incarnation. Chaitanya is Krishna, but He has taken the mode or role of Radharani to show us how to love Krishna. Thus He is addressed: "I offer my respectful obeisances unto the Supreme Lord who is absorbed in Radharani's thoughts."

This brings up the question of who Radharani is and what Radha-Krishna is. Actually Radha-Krishna is the exchange of love. This is not ordinary love; Krishna has immense potencies, of which three are principal: internal, external and marginal. In the internal potency there are three divisions: samvit, hladini and sandhini. The hladini potency is the pleasure potency. All living entities have this pleasure-seeking potency, for all beings are trying to have pleasure.

This is the very nature of the living entity. At present we are trying to enjoy our pleasure potency by means of the body in this material condition. By bodily contact we are attempting to derive pleasure from material sense objects. We should not think, however, that Krishna, who is always spiritual, tries to seek pleasure on this material plane like us. Krishna describes the material universe as a nonpermanent place full of miseries. Why, then, would He seek pleasure in the material form? He is the Supersoul, the supreme spirit, and His pleasure is beyond the material conception.

In order to learn how Krishna's pleasure can be obtained, we must read the Tenth Canto of Srimad-Bhagavatam in which Krishna's pleasure potency is displayed in His pastimes with Radharani and the damsels of Vraja. Unfortunately, unintelligent people turn at once to the sports of Krishna in the Dasama-skandha, the Tenth Canto. Krishna's embracing Radharani or His dancing with the cowherd girls in the rasa dance are generally not understood by ordinary men because they consider these pastimes in the light of mundane lust.

They incorrectly think that Krishna is like themselves and that He embraces the gopis just as an ordinary man embraces a young girl. Some people thus become interested in Krishna because they think that His religion allows indulgence in sex. This is not Krishna-bhakti, love of Krishna, but prakrta-sahajiya—materialistic lust.

In order to avoid such errors, we should understand what Radha-Krishna actually is. Radha and Krishna display their pastimes through Krishna's internal energy. The pleasure potency of Krishna's internal energy is a most difficult subject matter, and unless one understands what Krishna is, he cannot understand it. Krishna does not take any pleasure in this material world, but He has a pleasure potency. Because we are part and parcel of Krishna, the pleasure potency is within us also, but we are trying to exhibit that pleasure potency in matter.

Krishna, however, does not make such a vain attempt. The object of Krishna's pleasure potency is Radharani, and He exhibits His potency or His energy as Radharani and then engages in loving affairs with Her. In other words, Krishna does not take pleasure in this external energy but exhibits His internal energy, His pleasure potency, as Radharani. Thus Krishna manifests Himself as Radharani in order to exhibit His internal pleasure potency. Of the many extensions, expansions and incarnations of the Lord, this pleasure potency is the foremost and chief.

It is not that Radharani is separate from Krishna. Radharani is also Krishna, for there is no difference between the energy and the energetic. Without energy, there is no meaning to the energetic, and without the energetic, there is no energy. Similarly, without Radha there is no meaning to Krishna, and without Krishna, there is no meaning to Radha. Because of this, the Vaishnava philosophy first of all pays obeisances to and worships the internal pleasure potency of the Supreme Lord. Thus the Lord and His potency are always referred to as Radha-Krishna. Similarly, those who worship the name of Narayana first of all utter the name of Lakshmi, as Lakshmi-Narayana. Similarly, those who worship Lord Rama first of all utter the name of Sita. In any case—Sita-Rama, Radha-Krishna, Lakshmi-Narayana—the potency always comes first.

Radha and Krishna are one, and when Krishna desires to enjoy pleasure, He manifests Himself as Radharani. The spiritual exchange of love between Radha and Krishna is the actual display of the internal pleasure potency of Krishna. Although we speak of "when" Krishna desires, just when He did desire we cannot say. We only speak in this way because in conditional life we take it that everything has a beginning; however, in the absolute or spiritual life there is neither beginning nor end. Yet in order to understand that Radha and

Krishna are one and that They also become divided, the question "When?" automatically comes to mind. When Krishna desired to enjoy His pleasure potency, He manifested Himself in the separate form of Radharani, and when He wanted to understand Himself through the agency of Radha, He united with Radharani, and that unification is called Lord Chaitanya.

Why did Krishna assume the form of Chaitanya Mahaprabhu? It is explained that Krishna desired to know the glory of Radha's love. "Why is She so much in love with Me?" Krishna asked. "What is My special qualification that attracts Her so? And what is the actual way in which She loves Me?" It seems strange that Krishna, as the Supreme, should be attracted by anyone's love. We search after the love of a woman or a man because we are imperfect and lack something. The love of a woman, that potency and pleasure, is absent in man, and therefore a man wants a woman, but this is not the case with Krishna, who is full in Himself.

Thus Krishna expressed surprise: "Why am I attracted by Radharani? And when Radharani feels My love, what is She actually feeling?" In order to taste the essence of that loving affair, Krishna appeared just as the moon appears on the horizon of the sea. Just as the moon was produced by the churning of the sea, by the churning of spiritual love affairs the moon of Chaitanya Mahaprabhu appeared. Indeed, Chaitanya's complexion was golden, just like the moon. Although this is figurative language, it conveys the meaning behind the appearance of Chaitanya Mahaprabhu. The full significance of His appearance will be explained in later chapters.

The manifestations of the Supreme are also explained in Chaitanya-caritamrta. After offering respects to Lord Chaitanya, Krishnadasa Kaviraja next offers them to Nityananda. He explains that Nityananda is a manifestation of Sankarsana, who is the origin of the Maha-Vishnu. Krishna's first manifestation is as Balram and then Sankarsana, and after Sankarsana He is manifest as Pradyumna. In this way so many expansions take place.

Although there are many expansions, Lord Sri Krishna is the origin, as confirmed in Brahma-samhita. He is like the original candle from which many thousands and millions of candles are lit. Although any number of candles can be lighted, the original candle still retains its identity as the origin. In this way Krishna expands Himself into so many lights, and all these expansions are called Vishnu-tattva. Vishnu is a large light, and we are small lights, but all are expansions of Krishna.

When it is necessary to create the material universe, Vishnu expands Himself as the Maha-Vishnu. This Maha-Vishnu lies down on the Causal Ocean and breathes all the universes from His nostrils. Thus from the Maha-Vishnu and the Causal Ocean all the universes spring, and all these universes float in the Causal Ocean. In this regard there is the story of Vamana, who, when He took three steps, stuck His foot through the covering of the universe. Water from the Causal Ocean flowed through the hole which His foot made, and it is said that flow of water became the River Ganges. Therefore the Ganges is accepted as the most sacred water of Vishnu and is worshipped by all Hindus from the Himalayas down to the Bay of Bengal.

That Maha-Vishnu who lies on the Causal Ocean is actually an expansion of Balram, who is Krishna's first expansion, and, in the Vrndavana pastimes, is the brother of Krishna. In the maha-mantra Hare Krishna, Hare Krishna, Krishna Krishna, Hare Hare. Hare Rama, Hare Rama, Rama Rama, Hare Hare, the word Rama refers to Balram. Since Nityananda is an expansion of Balram, Rama also refers to Lord Nityananda. Thus Hare Krishna, Hare Rama addresses not only Krishna and Balram but Lord Chaitanya and Nityananda as well.

The subject matter of Chaitanya-caritamrta primarily deals with what is beyond this material creation. The cosmic material expansion is called maya because it has no eternal existence. Because it is sometimes manifested and sometimes not manifested, it is regarded as illusory. But beyond this temporary manifestation there is a higher nature, as indicated in Bhagwad Gita:

paras tasmat tu bhavo 'nyo
'vyakto 'vyaktat sanatanah
yah sa sarvesu bhutesu
nasyatsu na vinasyati

"Yet there is another nature, which is eternal and is transcendental to this manifested and unmanifested matter. It is supreme and is never annihilated. When all in this world is annihilated, that part remains as it is." (Bg. 8.20)

That supreme nature is beyond the manifested (vyaktah) and unmanifested (avyaktah). This superior nature which is beyond both creation and annihilation is the living force which is manifest in the bodies of all living entities. The body itself is composed of inferior nature, matter; but it is the superior nature that is moving the body. The symptom of that superior nature is consciousness. Thus in the spiritual world, where everything is composed of the superior nature, everything is conscious. In the material world inanimate objects are not conscious, but in the spiritual world this is not so. There a table is conscious, the land is conscious, the trees are conscious—everything is conscious.

It is not possible to imagine how far this material manifestation extends. In the material world everything is calculated by imagination or by some imperfect method, but Vedic literatures give information of what lies beyond the material universe. Those who believe in experimental knowledge may doubt the Vedic conclusions, for they cannot even calculate how far this universe is extended, nor can they reach far into the universe itself.

It is not possible to obtain information of anything beyond this material nature by experimental means. That which is beyond our power of conception is called acintya, inconceivable. It is useless to argue or speculate about what is inconceivable. If it is truly inconceivable, it is not subject to speculation or experimentation. Our energy is limited, and our sense perception is limited; therefore we must rely on the Vedic conclusions regarding that subject matter which is inconceivable. Knowledge of the superior nature must simply be accepted without argument. How is it possible to argue about something to which we have no access? The method for understanding transcendental subject matter is given by Lord Krishna Himself in Bhagwad Gita, where Krishna tells Arjuna at the beginning of the Fourth Chapter:

imam vivasvate yogam
proktavan aham avyayam
vivasvan manave praha
manur iksvakave 'bravit

"I instructed this imperishable science of yoga to the sun-god, Vivasvan, and Vivasvan instructed it to Manu, the father of mankind, and Manu in turn instructed it to Iksvaku." (Bg. 4.1)

This is the method of parampara, or disciplic succession. Similarly, in Srimad-Bhagavatam Krishna imparted knowledge into the heart of Brahma, the first created creature within the universe. Brahma imparted those lessons to his disciple, Narada, and Narada imparted that knowledge to his disciple, Vyasadeva. Vyasadeva imparted it to Madhvacarya, and from Madhvacarya the knowledge comes down to Madhavendra Puri, to Isvara Puri and from him to Chaitanya Mahaprabhu.

One may ask that if Chaitanya Mahaprabhu is Krishna Himself, then why did He need a spiritual master? Of course He did not need a spiritual master, but because He was playing the role of acharya (one who teaches by example), He accepted a spiritual master. Even Krishna Himself accepted a spiritual master, for that is the system. In this way the Lord sets the example for men. We should not think, however, that the Lord takes a spiritual master because He is in want of knowledge. He is simply stressing the importance of accepting the disciplic succession.

The knowledge of that disciplic succession actually comes from the Lord Himself, and if the knowledge descends unbroken, it is perfect. Although we may not be in touch with the original personality who first imparted the knowledge, we may receive the same knowledge through this process of transmission. In Srimad-Bhagavatam, it is stated that Krishna, the Absolute Truth, the Personality of Godhead, transmitted transcendental knowledge into the heart of Brahma. This then is one way knowledge is received—through the heart. Thus there are two processes by which one may receive knowledge: One depends upon the Supreme Personality of Godhead, who is situated as the Supersoul within the heart of all living entities, and the other depends upon the guru or spiritual master, who is an expansion of Krishna. Thus Krishna transmits information both from within and from without. We simply have to receive it. If knowledge is received in this way, it doesn't matter whether it is inconceivable or not.

In Srimad-Bhagavatam there is a great deal of information given about the Vaikuntha planetary systems which are beyond the material universe. Similarly, a great deal of inconceivable information is given in Chaitanya-caritamrta. Any attempt to arrive at this information through experimental knowledge is not possible.

The knowledge simply has to be accepted. According to the Vedic method, sabda, or transcendental sound, is regarded as evidence. Sound is very important in Vedic understanding, for, if it is pure, it is accepted as authoritative. Even in the material world we accept a great deal of information which is sent thousands of miles by telephone or radio. In this way we also accept sound as evidence in our daily lives. Although we cannot see the informant, we accept his information as valid on the basis of sound. Sound vibration then is very important in the transmission of Vedic knowledge.

The Vedas inform us that beyond this cosmic manifestation there are extensive planets and the spiritual sky. This material manifestation is regarded as only a small portion of the total creation. The material manifestation includes not only this universe but innumerable others as well, but all the material universes combined comprise only one fraction of the total creation. The majority of the creation is situated in the spiritual sky. In that sky innumerable planets float, and these are called Vaikunthalokas. In every Vaikunthaloka Narayana presides in the form of His four-armed expansions: Sankarsana, Pradyumna, Aniruddha and Vasudeva.

As stated before, the material universes are manifested by the Lord in the form of Maha-Vishnu. Just as a husband and wife combine to beget offspring, the Maha-Vishnu combines with His wife Maya, or material nature. This is also confirmed in the Bhagwad Gita where Krishna states:

sarva-yonisu kaunteya
murtayah sambhavanti yah
tasam brahma mahad yonir
aham bija-pradah pita

"It should be understood that all species of life, O son of Kunti, are made possible by birth in this material nature, and that I am the seed-giving father." (Bg. 14.4)

Vishnu impregnated Maya or material nature simply by glancing at her. This is the spiritual method. Materially we are limited to impregnate by only one particular part of our body, but the Supreme Lord, Krishna or Maha-Vishnu, can impregnate any part by any part. Simply by glancing the Lord can conceive countless living entities in the womb of material nature. The Brahma-samhita also confirms that the spiritual body of the Supreme Lord is so powerful that any part of that body can perform the functions of any other part. We can only touch with our hands or skin, but Krishna can touch just by glancing. We can only see with our eyes, we cannot touch or smell with them.

Krishna, however, can smell and also eat with His eyes. When foodstuffs are offered to Krishna we don't see Him eating, but He eats simply by glancing at the food. We cannot imagine how things work in the spiritual world where everything is spiritual. It is not that Krishna does not eat or that we imagine that He eats; He actually eats, but His eating is different from ours. Our eating process will be similar to His when we are completely on the spiritual platform. On that platform every part of the body can act on behalf of any other part.

Vishnu does not require anything in order to create. He does not require the goddess Lakshmi in order to give birth to Brahma, for Brahma is born from a lotus flower which grows from the navel of Vishnu. The goddess Lakshmi sits at the feet of Vishnu and serves Him. In this material world sex is required to produce children, but in the spiritual world one can produce as many children as he likes without having to take help from his wife. Because we have no experience with spiritual energy, we think that Brahma's birth from the navel of Vishnu is simply a fictional story. We are not aware that spiritual energy is so powerful that it can do anything and everything. Material energy is dependent on certain laws, but spiritual energy is fully independent.

Brahma is born from the navel of Garbhodakasayi Vishnu, who is but a partial manifestation of the Maha-Vishnu. Countless universes reside like seeds within the skin pores of the Maha-Vishnu, and when He exhales, they all are manifest. In the material world we have no experience of such a thing, but we do experience a perverted reflection in the phenomenon of perspiration. We cannot imagine, however, the duration of one breath of the Maha-Vishnu, for within one breath all of the universes are created and annihilated.

Lord Brahma only lives for the duration of one breath, and according to our time scale 4,320,000,000 years constitute only twelve hours of Brahma, and Brahma lives one hundred of his years. Yet the whole life of Brahma is contained within one breath of the Maha-Vishnu. Thus it is not possible for us to imagine the breathing power of the Supreme Lord. That Maha-Vishnu is but a partial manifestation of Krishna.

Thus Krishnadasa Kaviraja Gosvami discusses Lord Chaitanya Mahaprabhu as Sri Krishna Himself, the Supreme Personality of Godhead, and Lord Nityananda as Balram, the first expansion of Krishna. Advaitacarya, another principal disciple of Lord Chaitanya Mahaprabhu's, is accepted as an expansion of the Maha-Vishnu. Thus Advaitacarya is also the Lord, or, more precisely, an expansion of the Lord. The word advaita means nondual, and his name is such because he is nondifferent from the Supreme Lord. He is also called acharya, teacher, because he disseminated Krishna consciousness.

In this way he is just like Chaitanya Mahaprabhu. Although Chaitanya is Sri Krishna Himself, He appears as a devotee to teach people in general how to love Krishna. Similarly, Advaitacarya appeared just to distribute the knowledge of Krishna consciousness. Thus he is also the Lord incarnated as a devotee. Krishna is manifested in five different expansions, and He and all of His associates appear as devotees of the Supreme Lord in the form of Sri Krishna Chaitanya, Nityananda, Advaitacarya, Gadadhara, Srivasa and others. In ali cases, Chaitanya Mahaprabhu is the source of energy for ali His devotees.

Since this is the case, if we take shelter of Chaitanya Mahaprabhu for the successful execution of Krishna consciousness, we are sure to make progress. One devotional song by Narottama dasa Thakura states: "My dear Lord Chaitanya, please have mercy upon me. There is no one who is as merciful as You. My plea is most urgent because Your mission is to deliver fallen souls, and no one is more fallen than l. I beg priority."

The author of Chaitanya-caritamrta, Krishnadasa Kaviraja Gosvami, was an inhabitant of Vrndavana and a great devotee. He had been living with his family in Katwa, a small town in the district of Burdwan in Bengal. His family also worshipped Radha-Krishna, and once when there was some misunderstanding amongst his family about devotional service, Krishnadasa Kaviraja was advised by Nityananda Prabhu in a dream to leave home and go to Vrndavana. Although he was very old, he started out that very night and went to live in Vrndavana. While he was there, he met some of the Goshwamis, principal disciples of Lord Chaitanya Mahaprabhu. He was requested to write Chaitanya-caritamrta by the devotees of Vrndavana. Although he began this work at a very old age, by the grace of Lord Chaitanya he finished it. Today it remains the most authoritative book on Chaitanya's philosophy and life.

When Krishnadasa Kaviraja Gosvami was living in Vrndavana, there were not very many temples. At that time Madana-mohana, Govindaji and Gopinatha were the three principal temples. As a resident of Vrndavana, he offered his respects to the Deities in these temples and requested God's favour: "My progress in spiritual life is very slow, so l'm asking Your help." In Chaitanya-caritamrta, Krishnadasa first offers his obeisances to Madana-mohana vigraha, the Deity who can help us progress in Krishna consciousness. In the execution of Krishna consciousness, our first business is to know Krishna and our relationship with Him. To know Krishna is to know one's self, and to know one's self is to know one's relationship with Krishna. Since this relationship can be learned by worshipping Madana-mohana vigraha, Krishnadasa Kaviraja Gosvami first establishes his relationship with Him.

When this is established, Krishnadasa begins to worship the functional Deity, Govinda. Govinda resides eternally in Vrndavana. In the spiritual world of Vrndavana the buildings are made of touchstone, the cows are known as surabhi cows, givers of abundant milk, and the trees are known as wish-fulfilling trees, for they yield whatever one desires. In Vrndavana Krishna herds the surabhi cows, and He is worshipped by hundreds and thousands of gopis, cowherd girls, who are all goddesses of fortune. When Krishna descends to the material world, this same Vrndavana descends just as an entourage accompanies an important personage.

Because when Krishna comes, His land also comes, Vrndavana is not considered to exist in the material world. Therefore devotees take shelter of the Vrndavana in India, for it is considered to be a replica of the original Vrndavana. Although one may complain that no kalpa-vrksa, wish-fulfilling trees, exist there, when the Gosvamis were there, the kalpa-vrksa were present. It is not that one can simply go to such a tree and make demands; one must first become a devotee. The Gosvamis would live under a tree for one night only, and the trees would satisfy all their desires. For the common man this may all seem very wonderful, but as one makes progress in devotional service, all this can be realized.

Vrndavana is actually experienced as it is by persons who have stopped trying to derive pleasure from material enjoyment. "When will my mind become cleansed of all hankering for material enjoyment so I will be able to see Vrndavana?" one great devotee asks. The more Krishna conscious we become and the more we advance, the more everything is revealed as spiritual. Thus Krishnadasa Kaviraja Gosvami considered Vrndavana in India to be as good as the Vrndavana in the spiritual sky, and in Chaitanya-caritamrta he describes Radharani and Krishna as seated beneath a wish-fulfilling tree in Vrndavana on a throne decorated with valuable jewels. There Krishna's dear friends, the cowherd boys and the gopis, serve Radha and Krishna by singing, dancing, offering betel nuts and refreshments and decorating Their Lordships with flowers. Even today in India people decorate thrones and recreate this scene during the month of July. Generally at that time people go to Vrndavana to offer their respects to the Deities there.

Krishnadasa Kaviraja Gosvami maintains that the Radha and Krishna Deities show us how to serve Radha and Krishna. The Madana-mohana Deities simply establish that "I am Your eternal servant." With Govinda, however, there is actual acceptance of service, and therefore He is called the functional Deity. The Gopinatha Deity is Krishna as master and proprietor of the gopis. He attracted all the gopis, or cowherd girls, by the sound of His flute, and when they came, He danced with them. These activities are all described in the Tenth Canto of Srimad-Bhagavatam.

These gopis were childhood friends of Krishna, and they were all married, for in India the girls are married by the age of twelve. The boys, however, are not married before eighteen so Krishna, who was fifteen or sixteen at the time, was not married. Nonetheless He called these girls from their homes and invited them to dance with Him. That dance is called the rasa-lila dance, and it is the most elevated of all the Vrndavana pastimes. Krishna is therefore called Gopinatha because He is the beloved master of the gopis.

Krishnadasa Kaviraja Gosvami petitions the blessings of Lord Gopinatha. "May that Gopinatha, the master of the gopis, Krishna, bless you. May you become blessed by Gopinatha." Just as Krishna attracted the gopis by the sweet sound of His flute, the author of Chaitanya-caritamrta prays that He will also attract the reader's mind by His transcendental vibration. It is the purpose of this book, Teachings of Lord Chaitanya, to transmit the essence of that vibration in an easily readable summary study.

Teachings to Rupa Gosvami

Srila Rupa Gosvami, the younger brother of Sanatana Gosvami, went to Prayaga, the modern city of Allahabad, with his younger brother Vallabha. When the two brothers heard that Lord Sri Chaitanya Mahaprabhu was staying there, they both became very happy and went to see the Lord. At that time the Lord was on His way to visit the temple Bindumadhava. On the way to the temple, the Lord was chanting and dancing, and thousands of people were following Him. Some of these people were crying, and some were laughing. Some were dancing, and some were singing, and some were falling on the ground, offering obeisances to the Lord. In all cases, all of them were roaring the holy name of Krishna. It is said that in spite of being at the confluence of the rivers Ganges and Yamuna, Prayaga was never flooded until the appearance of Chaitanya Mahaprabhu, at which time the city was overflooded by love of Krishna.

The two brothers, Rupa Gosvami and Vallabha, stayed aloof in an uncrowded place and witnessed the great crowd and wonderful scene. When the Lord danced, He raised His arms and shouted, "Haribol! Haribol!" The people all about Him were astonished to see His wonderful activities. After visiting the temple, the Lord accepted prasada (food offered to the Deity) at the house of a Deccanist (Southern) brahmana. While at the brahmana's home, the Lord was visited by Rupa Gosvami and Vallabha. From a distance the two brothers fell down on the ground to offer obeisances, and they chanted many Sanskrit verses from the scriptures. When the Lord saw Rupa Gosvami offering obeisances before Him, He became very pleased and asked him to get up. The Lord then informed Rupa Gosvami of the causeless mercy of Krishna upon him, for Krishna had just delivered him from a materialistic way of life based simply on pound-shilling-pence.

The Lord accepted the two brothers as His own devotees, and He cited one verse from the scriptures which stated that it is possible for a brahmana, who has studied the four Vedas, to not be accepted as a devotee of the Lord and that a pure devotee could come from a very low family and yet be accepted by Him. Then the Lord embraced the two brothers, and, out of His causeless mercy, touched their heads with His lotus feet. Being blessed in this way, the brothers offered prayers to the Lord in their own words. The prayers indicated that Lord Sri

Krishna Chaitanya Mahaprabhu was Krishna Himself, that He had assumed the form and fair complexion of Chaitanya and was thus known as Gauranga (the golden one), and that He was the most munificent incarnation of Krishna because He was distributing love of Krishna. Srila Rupa Gosvami also quoted one verse which was later found in the book Govinda-lilamrta (1.2):

yo 'jnana-mattam bhuvanam dayalur
ullaghayann apy akarot pramattam
svaprema-sampat-sudhayadbhuteham
sri-Krishna-Chaitanyam amum prapadye

"Let me surrender unto the lotus feet of Sri Krishna Chaitanya Mahaprabhu, who is the most merciful Personality of Godhead. He delivers those souls who are merged in ignorance and offers them the highest gift, love of Krishna, and thus makes them mad for Krishna consciousness."

After this incident, Vallabha Bhatta invited the Lord to go to the other side of the Ganges, and the Lord went. From that moment on, wherever the Lord went, Rupa Gosvami would follow Him and stay with Him. Because the Lord felt inconvenienced in crowded places, He asked Rupa Gosvami to accompany Him to a place on the banks of the Ganges known as Dasasvamedha-ghata. For ten days He instructed Rupa Gosvami about the truth of Krishna, the principles of devotional service and the transcendental relationships with Krishna. All of this was described in full detail so that in the future Rupa Gosvami could distribute this science of Krishna in his book Bhakti-rasamrta-sindhu. Indeed, Srila Rupa Gosvami described this incident in the first verse of Bhakti-rasamrta-sindhu, in which he speaks of the causeless mercy of the Lord upon him.

The Supreme Lord is cognizant and all-powerful, and by His causeless mercy He empowers a living entity to receive His mercy. Being under the spell of conditional life, people in general are averse to rendering devotional service and practicing Krishna consciousness.

In fact, most people are unaware of the principal teachings of Krishna consciousness regarding one's eternal relationship with the Supreme Personality of Godhead and the ultimate goal of life, which is to return home, back to Godhead. Nor are people aware of the process by which one can return to the spiritual world. Because these important subject matters are unknown to the conditioned soul, Lord Chaitanya, out of His causeless mercy, instructed Rupa Gosvami in the principles of devotional service. Later, for the good of the people in general, Rupa Gosvami distributed this information of the science of devotional service.

In the prologue to Bhakti-rasamrta-sindhu (1.1.2), Rupa Gosvami wrote the following:

hrdi yasya preranaya
pravartito 'ham varaka-rupo 'pi
tasya hareh pada-kamalam
vande Chaitanyadevasya

"I offer my respectful obeisances unto the lotus feet of the Supreme Personality of Godhead, known as Lord Chaitanyadeva, because due to His inspiration I feel the desire within my heart to write something about devotional service. For this reason I am engaged in writing this book on the science of devotion known as Bhakti-rasamrta-sindhu."

When Lord Chaitanya began His instructions to Rupa Gosvami, He first told him, "My dear Rupa, the science of devotional service is just like the great ocean, and it is not possible to show you all its length and breadth. However, I shall try to explain the nature of that ocean by taking just one drop out of it. In this way you can taste it and understand what that ocean of devotional service actually is.

The Lord then explained that within this brahmanda, or universe, there are innumerable living entities who, according to their own fruitive activities, are transmigrating from one species of life to another and from one planet to another. In this way their encagement in material existence has been continuing since time immemorial. In actuality, these living entities are atomic parts and parcels of the supreme spirit. It is said in the Srimad-Bhagavatam that the length and breadth of the individual soul is approximately 1.10,000th part of the tip of a hair—in other words, it is so small that it is invisible. This is also confirmed in the Svetasvatara Upanisad. In the Tenth Canto of Srimad-Bhagavatam, one of the four Kumaras, known as Sanandana, gave the following speech upon performing a great sacrifice: "O Supreme Truth! If the living entities were not infinitesimal sparks of the supreme spirit, each minute spark would be all-pervading and would not be controlled by a superior power.

But if the living entity is accepted as a minute part and parcel of the Supreme Lord he automatically becomes controlled by a supreme energy or power. The latter is his actual constitutional position, and if he remains in this position he can attain full freedom." (Bhag. 10.87.30) If one mistakenly considers his position to be equal to that of the Supreme Personality of Godhead, he becomes contaminated by the doctrine of nonduality, and his efforts in transcendental life are rendered ineffective.

Lord Chaitanya elaborated on these teachings of Srimad-Bhagavatam by pointing out that there are two kinds of living entities—the eternally liberated and the eternally conditioned. The eternally conditioned living entities can be divided into two types—moving and nonmoving. Those entities which cannot move—like trees, for example—remain in one place and are classified as nonmoving entities, and those that move—such as the birds and beasts—are called jangama (moving entities) and are further divided into three categories: those that fly in the sky, those that swim in the water, and those that walk on land.

Out of the many millions and trillions of living entities on land, human beings comprise only a small portion. Out of that small number of human beings, most are totally ignorant of spiritual life, are unclean in their habits and have no faith in the existence of the Supreme Personality of Godhead. In short, most human beings live like animals. These can actually be deducted from the number of human beings that comprise human or civilized life. It is very difficult to find a few human beings who believe in the scriptures and the existence of God, or, for that matter, in proper behaviour. Those who do believe in the value of these things are known as arya, a word denoting those who believe in advancing in spiritual life. Out of those who believe in the value of the scriptures and the advancement of human civilization, there are two classes—the righteous and the unrighteous.

Those who are righteous generally execute fruitive activities in order to derive some good result for sense gratification. Out of many such persons who engage in righteous activities for sense gratification, only a few come to know about the Absolute Truth. These are called jnanis, empiric philosophers. Out of many hundreds and thousands of such empiric philosophers, only a handful actually attain liberation. When one is liberated, he theoretically understands that the living entity is not composed of material elements but is spirit soul, distinct from matter. Simply by theoretically understanding this doctrine, one can be called liberated, but actually a mukta, or liberated soul, is he who understands his constitutional position as an eternal servant of the Lord. Such liberated souls engage with faith and devotion in the service of the Lord, and they are called Krishna-bhaktas, or Krishna conscious persons.

Krishna-bhaktas are free from all material desires. Those who are liberated theoretically by knowing simply that the living entity is not material may still have desires, although they may be technically classified amongst liberated souls. Their main desire is to become one with the Supreme Personality of Godhead. Generally such persons are very much attached to Vedic rituals and righteous activities, performing them in order to enjoy material prosperity. Even when some of them transcend material enjoyment, they still try to enjoy the spiritual world by merging into the existence of the Supreme Lord.

Some of them are also desirous of attaining mystic powers through the execution of yoga. As long as these desires are within one's heart, he cannot understand the nature of pure devotional service. When one is constantly being agitated by such desires, he is not peaceful. Indeed, as long as there is any desire for material perfection at all, one cannot be at peace. Since the devotees of Lord Krishna do not desire anything material, they are the only peaceful persons within this material world. This is confirmed in Srimad-Bhagavatam:

muktanam api siddhanam
narayana-parayanah
sudurlabhah prasantatma
kotisv api mahamune

"O great sage, out of many millions of liberated persons and persons who have achieved success in mystic yoga, one who is completely devoted to the Supreme Personality of Godhead and who is filled with peace is very hard to find." (Bhag. 6.14.5)

In this way Lord Chaitanya explained that of the many thousands and millions of living entities wandering in the material world, one who by the grace of Lord Krishna and the spiritual master gets the seed of devotional service is very rare and fortunate. A pious or religious man is generally inclined to worship deities in various temples, but if by chance, even without his knowledge, he offers his obeisances to Lord Vishnu or receives the favour of a Vaishnava, a devotee of the Lord, he at that time acquires the asset necessary to approach the Supreme Personality of Godhead.

This is clearly understood from the life story of the great sage Narada, which is related in Srimad-Bhagavatam. By serving Vaishnavas in his previous life, Narada was favoured by the devotees of the Lord and became a great sage. Indeed, amongst sages Narada Muni is considered to be the greatest.

Vaishnavas, or devotees, are usually very compassionate upon conditioned souls. Without even being invited, a devotee will go from door to door to enlighten people and to bring them out of the darkness of nescience by injecting knowledge of the living entity's constitutional position as a servant of Lord Krishna. Such devotees are empowered by the Lord to distribute devotional consciousness, or Krishna consciousness, to the people in general. They are known as authorized spiritual masters, and it is by their mercy that a conditioned soul gets the seed of devotional service. The causeless mercy of the Supreme Personality of Godhead is first appreciated when one comes in touch with a bona fide spiritual master who can bring the conditioned soul to the highest position of devotional life. Therefore Lord Chaitanya said that by the mercy of the spiritual master one can achieve the causeless mercy of the Lord, and by the mercy of the Supreme Personality of Godhead, one can attain the mercy of the bona fide spiritual master.

Thus by the mercy of the spiritual master and Krishna, one receives the seed of devotional service. He has only to sow the seed in the field of his heart, just as a gardener sows the seed of a valuable tree. After sowing this seed, one has to water it in the form of chanting and hearing the holy name of the Supreme Lord or by taking part in discussions about the science of devotional service in a society of pure devotees. When the plant of devotional service sprouts up from the seed of devotion, it begins to grow freely. When it is full grown, it surpasses the length and breadth of this universe and enters into the transcendental atmosphere, where everything is bathed in the effulgence of the brahmajyoti.

The plant even penetrates this brahmajyoti and gradually enters the planet known as Goloka Vrndavana. There the plant takes shelter at the lotus feet of Krishna. That is the ultimate goal of devotional service. After attaining this position, the plant produces fruit, which is known as the fruit of love of Godhead. However, it is necessary for the devotee, or transcendental gardener, to pour water on the plant daily by chanting and hearing. Unless one waters the plant by chanting and hearing, there is every chance that it will dry up.

Lord Chaitanya pointed out to Rupa Gosvami that there was a certain danger to be encountered while watering the root of the devotional plant. After the plant has grown some bit, an animal may come and either eat or destroy it. When green leaves of a plant are taken by some animal, the plant generally dies.

The most dangerous animal is considered a mad elephant, for if a mad elephant enters into a garden, it causes tremendous damage to plants and trees. An offense to a pure devotee of the Lord is called Vaishnavaparadha, the mad elephant offense. In the discharge of devotional service, an offense to the feet of a pure devotee can create havoc. Thus one has to defend the plant of bhakti by tending it properly and taking care not to commit offenses. If one is cautious, the plant can properly thrive.

There are ten principal offenses which can be committed against the holy name. The first is to blaspheme the great devotees who have tried to spread the glories of the holy name throughout the world. The holy name of Krishna is nondifferent from Krishna, and one who attempts to spread the holy names throughout the world is beloved of Him. Krishna Himself does not tolerate offenses against His pure devotees. The second offense is to deny that Lord Vishnu is the Absolute Truth. There is no difference between His name, quality, form, pastimes and activities, and one who sees a difference is considered an offender. The Lord is Supreme, and no one is equal to or greater than Him. Consequently if one thinks that the Lord's names are nondifferent from the names of demigods, he offends. The Supreme Lord and the demigods should never be considered on the same level.

The third offense is to consider the bona fide spiritual master to be a common man. The fourth offense is to blaspheme Vedic literature and authorized scriptures like the Puranas. The fifth offense is to consider the glories attributed to the holy names to be exaggerations. The sixth offense is to concoct perverted theories about the holy name. The seventh offense is to commit sinful activities on the strength of chanting the holy name. It is understood that by chanting the holy names one is free from sinful reactions, but this does not mean that one should act sinfully on the strength of chanting. That is the greatest offense.

The eighth offense is to consider that religious rituals, austerity, sacrifices or other forms of renunciation are equal to chanting the holy name. Chanting the holy name is as good as associating with the Supreme Personality of Godhead. Pious activities are only means to approach the Supreme Personality of Godhead, and they can even be performed for some material reason. The ninth offense is to preach the glories of the holy name of God to a faithless person who is not interested in hearing them. The tenth and last offense is to maintain material attachment even after hearing and chanting the holy names of God. The idea is that by chanting the holy name without offense, one can obtain elevation to the liberated platform. On the liberated platform one is freed from all material attachment. Thus if one chants the holy names and still has material attachments, he must be committing some offense.

There are also other factors which disturb the plant of devotional service. Along with this plant the weeds of material desires also grow. When a person advances in bhakti, it is natural that many persons will come to him requesting to become disciples and will offer him some material gains. If one is attracted by a large number of disciples and material conveniences offered by these disciples and forgets his duty as a bona fide master, the growth of the plant will be impeded. Simply by taking advantage of material conveniences one may become addicted to enjoying material comforts.

It is also considered to be disadvantageous to desire liberation. The only desire should be the desire to render service. Neglect of restrictions and prohibitions is also disadvantageous. The prohibitions are mentioned in the authorized scriptures: One should not indulge in illicit sex life, intoxication, meat eating or gambling. These things are forbidden to one who is attempting devotional service. If one does not follow these principles strictly, there may be a severe disturbance in the discharge of devotional service.

If one is not particularly careful, even by watering the plant of devotional service, unnecessary weeds will grow and hamper progress. The idea is that when one waters a garden, not only does the desired plant grow more rapidly, but the unwanted plants grow also. If the gardener does not see these impediments and take them out, they will overcome and choke the plant of devotion. If, however, one is careful to guard against the growth of unwanted plants, the plant of devotion grows luxuriantly and reaches the ultimate goal, Goloka Vrndavana. When the living entity engaged in devotional service relishes the fruit of love of Godhead, he forgets all religious ritual and improvements in his economic condition. He no longer desires to satisfy his senses, and he no longer desires to become one with the Supreme Lord by merging into His effulgence.

There are many phases of spiritual knowledge and transcendental bliss. On one platform are the ritualistic sacrifices recommended in the Vedas, the execution of austerities and pious duties, and the practice of mystic yoga. These all reward different results to their performer. The rewards of these practices, however, appear to be very glittering as long as one is not elevated to the transcendental loving service of the Lord. Love of God is dormant in everyone, and it can be awakened from its dormant position by the execution of pure devotional service, just as a person bitten by a serpent can be awakened by ammonia.

After speaking in this way about devotional service, Lord Chaitanya begins to describe devotional service and its symptoms to Rupa Gosvami. He explains that in pure devotional service there can be no desire other than the desire to advance in Krishna consciousness. In Krishna consciousness there is no scope for worshipping any demigod or any other form of Krishna, nor is there room for indulgence in speculative empiric philosophy, nor indulgence in fruitive activities. One should be free from all these contaminations. A devotee should accept only those things that are favourable to keep his body and soul together and should reject those things that increase the demands of the body.

Only the bare necessities for bodily maintenance should be accepted. By minimizing bodily necessities, one can primarily devote his time to the cultivation of Krishna consciousness through the chanting of the holy names of God. Pure devotional service means engaging all the senses of the body in the service of the Lord. At the present moment, our senses are all designated because the body is designated. Consequently we think that this body belongs to a particular society or a particular country or a particular family. In this way the body is bound by so many designations. Similarly, the senses belong to the body, and when the body is subject to such designations, the senses are also.

Thus the senses engage themselves on behalf of family, society, nation and so on. When they are so engaged, they cannot cultivate Krishna consciousness. The senses must be purified, and this is possible when one purely understands that he belongs to Krishna and that his life belongs to Krishna. The devotee should see his identity as an eternal servant of Krishna. In this way one can engage his senses in the service of the Lord. Such engagement is called pure devotional service.

A pure devotee accepts the transcendental loving service of the Lord but rejects ali kinds of liberation for his personal sense gratification. In Srimad-Bhagavatam (3.29.11-13) Lord Kapila explains that as soon as a pure devotee hears the glories and transcendental qualities of the Supreme Personality of Godhead, who is seated in everyone's heart, his mind immediately flows toward the Lord, just as the waters of the Ganges flow toward the sea. Such spontaneous attraction to the service of the Supreme Personality of Godhead is most important to pure devotional service. Devotional service is pure when one engages in the service of the Supreme Lord without any motive and without being hampered with material impediments.

The pure devotee does not desire to live on the same planet with the Supreme Lord, nor does he desire the same opulence as the Lord, nor does he desire to have the same form as the Lord, nor to live with Him side by side, nor to merge into His existence, etc. Even if the devotee were offered such rewards by the Lord, he would reject them.

The point is that a devotee is so much absorbed in the transcendental loving service of the Lord that he has no time to think of any benefit beyond his immediate engagement. Just as an ordinary materialistic businessman thinks of nothing else when he is absorbed in his business, a pure devotee, when engaged in the service of the Lord, does not think of anything beyond that engagement.

If one is so absorbed in the rendering of service, he can be understood to be elevated to the highest position of bhakti. By such transcendental loving service alone can one surpass the influence of maya and relish pure love of Godhead. As long as one desires material benefit or liberation, which are called the two witches of allurement, he cannot relish the taste of transcendental loving service to the Supreme Lord. There are three stages of devotional service: The first is the beginning stage of cultivation, the second is the realization of service, and the third, the supreme stage, is the attainment of love of Godhead.

There are nine different methods of cultivating devotional service—such as hearing, chanting, remembering, etc.—and all these processes are employed in the first stage. If one is engaged in chanting and hearing with devotion and faith, his material misgivings gradually become vanquished. As his faith in devotional service gradually increases, he becomes assured of a higher perfectional position. In this way one can become firmly fixed in devotion, increase his taste for it, become attached and feel ecstasy. This ecstasy occurs in the preliminary stage of love of Godhead. Attainment of ecstasy is produced by execution of devotional service. When one continues the process of hearing and chanting, attachment grows and assumes the name of love of Godhead.

When one attains the third stage of transcendental love of God, there occur further developments known as transcendental affection, emotion, ecstasy, and extreme and intense attachment. These are technically known by the terms raga, anuraga, bhava and mahabhava. The progress from one stage to another can be compared to the thickening of sugar candy juice. In the first stage sugar candy juice is like a thin liquid. When, by evaporation, it becomes thicker and thicker, it turns into molasses. Finally it turns into granules and becomes sugar, rock candy and so on. Just as liquid sugar juice progresses from one stage to another, similarly transcendental love for the Supreme Lord develops by stages.

When one actually becomes situated on the transcendental platform, he becomes steady. Unless one is so situated, his position may not be steady and he may fall down. When one is actually situated transcendentally, there is no fear of falling down. This stage of understanding is technically called sthayi-bhava. There are even stages beyond this position, and they are known as vibhava, anubhava, sattvika and vyabhicari. After one attains these, there is actually an exchange of rasa, or transcendental activity with the Supreme Lord. This exchange in loving reciprocation between the lover and the beloved is generally called Krishna-bhakti-rasa. It should be noted that the transcendental loving exchanges stand on the steadfast position of sthayi-bhava, as explained before. The basic principle of vibhava is sthayi-bhava, and all other activities are auxiliary for the development of transcendental love.

The ecstasy of transcendental love has two components—the context and the cause of the excitement. The context is also divided into two parts—the subject and the object. The exchange of devotional service is the subject, and Krishna is the object. The transcendental qualities are the causes of excitement. This means that the transcendental qualities of Krishna excite the devotee to serve Him. The impersonal (Mayavadi) philosophers say that the Absolute Truth has no specific qualities, but the Vaishnava philosophers say that the Absolute Truth is described as nirguna (without qualities) because He has no material qualities. This is not to say that He does not have spiritual qualities. Indeed, the Lord's spiritual qualities are so great and so enchanting that they can even attract a liberated person. This is explained in the atmarama verse of the Srimad-Bhagavatam where it is said that those who are already situated on the platform of self-realization are attracted by the transcendental qualities of Krishna. This means that Krishna's qualities are not material but pure and transcendental.

The higher stage of ecstasy can be characterized by the following thirteen transcendental activities: (1) dancing, (2) rolling on the floor, (3) singing, (4) clapping, (5) bristling of the hairs of the body, (6) thundering, (7) yawning, (8) breathing heavily, (9) forgetting social conventions, (10) salivating, (11) laughing, (12) aching, (13) coughing. All these symptoms are not awakened simultaneously; they act according to the exchange of transcendental relationships. Sometimes one symptom is prominent, and at another time another is prominent.

The transcendental rasas, or relationships, can be divided into five. The initial stage is called santa-rati, wherein one who is liberated from material contamination appreciates the greatness of the Supreme Personality of Godhead. One who attains this stage does not exactly engage in the transcendental loving service of the Lord, for this is a neutral stage. In the second stage, which is called dasya-rati, a person appreciates his position as being everlastingly subordinate to the Supreme Lord, and he understands that he is eternally dependent on the causeless mercy of the Supreme Person. At that same time there is an awakening of natural affection, such as is felt by a son who grows up and begins to appreciate his father's benedictions.

At this stage the living entity wants to serve the Supreme Lord instead of serving maya, illusion. In the third stage, called sakhya-rati, transcendental love is developed, and one associates with the Supreme on an equal level of love and respect. As this stage is further developed, there is joking and such relaxed exchanges as laughing and so on. On this level there are fraternal exchanges with the Supreme Person, and one is free from all bondage. At this stage one practically forgets his inferior position as a living entity, but at the same time he has the greatest respect for the Supreme Person.

In the fourth stage, called vatsalya-rati, the fraternal affection evinced in the preceding stage develops into paternal affection. At this time the living entity tries to be the parent of God. Instead of worshipping the Lord, the living entity, as a parent of the Supreme, becomes an object of worship for the Supreme Person. At this stage the Lord depends on the mercy of His pure devotee and puts Himself under the control of the devotee to be raised. The devotee in this stage attains the position wherein he can embrace the Supreme Lord and even kiss His head. In the fifth stage, called madhura-rati, there is an actual transcendental exchange of conjugal love between the lover and the beloved. It is at this stage that Krishna and the damsels of Vraja glanced at one another, for on this platform there is an exchange of loving glances, motions of the eyes, pleasant words, attractive smiles, etc.

Besides these five primary rasas, or relationships, there are seven secondary rasas which consist of laughing, having wonderful visions, entering into a chivalrous relationship, experiencing pity, feeling anger and experiencing ghastliness and devastation. For example, Bhishma related to Krishna as a warrior in the chivalrous rasa. Hiranyakasipu, however, experienced an exchange of the ghastly and devastating rasa. The five primary rasas constantly remain within the heart of the pure devotee, and the seven secondary rasas sometimes appear and disappear to enrich the flavours and tastes of the primary ones. After enriching the primary rasas, they disappear.

Examples of santa-bhaktas, or devotees in the neutral stage, are the nine yogis named Kavi, Havi, Antariksa, Prabuddha, Pippalayana, Avirhotra, Dravida or Drumila, Camasa and Karabhajana. The four Kumaras (Sanaka, Sanandana, Sanatkumara and Sanatana) are also examples of this stage. Examples of devotees in the second stage, the dasya stage of servitorship, are Raktaka, Citraka and Patraka in the Gokula rasa. These all function as servants of Krishna.

In Dvaraka there is Daruka, and in the Vaikuntha planets there are Hanuman and others. Devotees in the third stage, the stage of friendship, are Sridama in Vrndavana and Bhima and Arjuna in Dvaraka and on the Battlefield of Kurukshetra. There are many others also. As far as those relating to Krishna in paternal love, they include devotees like Yashoda and Maharaja Nanda—that is, Krishna's mother, father, uncle and similar relatives. In conjugal love there are the damsels of Vraja, Vrndavana, and the queens and goddesses of fortune in Dvaraka. No one can count the vast number of devotees in this rasa.

Attachment to Krishna can also be broken down into two categories. On one platform there is attachment with awe and veneration. This type of attachment might be characterized by a certain lack of freedom, and it is exhibited in Mathura and in the Vaikuntha planets. In these abodes of the Lord, the spirit of transcendental loving service is restricted. However, in Gokula Vrndavana, love is freely exchanged, and although the cowherd boys and damsels of Vrndavana know that Krishna is the Supreme Personality of Godhead, they do not show awe and veneration because of the great intimacy of their relationship with Him. In the five principal transcendental relationships, awe and veneration are sometimes impediments obscuring the Lord's actual greatness and sometimes they actually impede one's service to the Lord. When there is friendship, paternal affection and conjugal love, however such awe and veneration are reduced. For instance, when Krishna appeared as the son of Vasudeva and Devaki, His parents prayed to the Lord with awe and veneration because they understood that the Supreme Lord Krishna or Vishnu had appeared before them as their little child.

This is confirmed in Srimad-Bhagavatam (10.44.51). Although the Supreme Lord was present as their child, Devaki and Vasudeva began to pray to Him. Similarly, when Arjuna saw the universal form of the Lord, he was so afraid that he begged pardon for his dealings with Krishna as an intimate friend. As a friend, Arjuna often behaved unceremoniously with the Lord, and upon seeing the awesome universal form, Arjuna said:

sakheti matva prasabham yad uktam
he Krishna he yadava he sakheti
ajanata mahimanam tavedam
maya pramadat pranayena vapi

yac cavahasartham asatkrto 'si
vihara-sayyasana-bhojanesu
eko 'thavapy achyuta tat-samaksam
tat ksamaye tvam aham aprameyam

"I have in the past addressed You as 'O Krishna,' 'O Yadava,' 'O my friend,' without knowing Your glories. Please forgive whatever I may have done in madness or in love. I have dishonoured You many times while relaxing or while lying on the same bed or eating together, sometimes alone and sometimes in front of many friends. Please excuse me for all my offenses." (Bg. 11.41-42)

Similarly, when Krishna was playing jokes on Rukmini, she feared that Krishna might leave her and became so perturbed that she dropped the fan with which she was fanning Him and fainted, falling unconscious on the floor. As far as Yashoda, Krishna's mother in Vrndavana, is concerned, it is stated in Srimad-Bhagavatam (Bhag. 10.8.45):

trayya copanisadbhis ca
sankhya-yogais ca satvataih

upagiyamana-mahatmyam
harim samanyatatmajam

The Personality of Godhead, who is worshipped by all the Vedas and Upanishads, as well as by the sankhya system of philosophy and all authorized scriptures, was considered to be born in her womb. It is also stated (Bhag. 10.9.12) that Mother Yashoda bound the child Krishna with a rope, as if He were an ordinary son born of her body. Similarly, there are other descriptions of Krishna's being treated as an ordinary person (Bhag. 10.18.24). Indeed, when He was defeated in games with His friends, the cowherd boys, Krishna would carry them—notably Sridama—on His shoulders.

Regarding the dealings of the gopis with Sri Krishna in Vrndavana, it is described (Bhag. 10.30.36-40) that when Sri Krishna took Srimati Radhika alone from the rasa dance, She thought that Krishna had left all the other gopis. Although they were all equally beautiful, He satisfied Her in this way, and She began to think proudly, "My dear Lord Krishna has left the beautiful gopis, and He is satisfied with Me alone." In the forest, She told Krishna, "My dear Krishna, I am unable to move anymore. Now if You like You can take Me wherever You desire." Krishna replied, "Come lean against My shoulder," and as soon as He said this, He disappeared, whereupon Srimati Radhika repined greatly.

When Krishna disappeared from the scene of the rasa dance, all the gopis began to repent, saying, "Dear Krishna! We have come here and have left aside our husbands, sons, relatives, brothers and friends! Neglecting their advice, we have come to You, and You best know the reason for our coming here. You know that we have come because we are captivated by the sweet sound of Your flute. But You are so cunning that in the dead of night You have left girls and women like us! This is not very good for You."

The word sama means controlling the mind and keeping it from being diverted in various ways by fixing it on the Supreme Personality of Godhead. When one's mind is fixed on the Supreme Lord, he is known to be situated on the sama platform. On that platform the devotee understands that

Krishna is the basic principle behind everything that is within one's experience. This is also explained in Bhagwad Gita (Bg. 7.19). Such a person can understand that Krishna is present in everything and is distributed all over the cosmic manifestation.

Although everything is under the control of the Supreme Lord and is situated in His energy, everything is nonetheless different from Krishna in His personal form. It is also stated in Bhakti-rasamrta-sindhu that one who understands this, whose intelligence is fixed on Krishna, has attained the platform of sama. Moreover, the Supreme Personality of Godhead says: samo mannisthata buddheh: Unless one is elevated to the platform of santa-rati, he cannot be fixed in knowledge of the greatness of Krishna or of the diffusion of His different energies, which are the cause of all manifestations. This same point is explained in Srimad-Bhagavatam (Bhag. 11.19.36):

samo mannisthata buddher
dama indriya-samyamah
titiksa duhkha-sammarso
jihvopastha jayo dhrtih

Stability of mind can be achieved by one who has concluded that the Supreme Personality of Godhead is the original source of everything. And when one can control his senses, that is called sama. When one is ready to tolerate all kinds of sufferings in order to control the senses and keep the mind steady, that is called titiksa, or tolerance. And when one can control the urges of the tongue and genitals, that is called dhrtih. From dhrtih, one becomes dhira, pacified. A pacified person is never disturbed by the urges of the tongue and the genitals.

If one can fix his mind on Krishna without deviation, he can attain a steadfast position in Krishna consciousness, santa-rasa. When one attains santa-rasa, unflinching faith in Krishna is established, and all material desires cease. These specific characteristics of santa-rasa—unflinching faith in Krishna and cessation of all desires which are not connected with Krishna—are common to all other rasas as well, just as sound is generally present in all other elements (air, fire, water and earth) because it is produced from the sky. Similarly, these two characteristics of santa-rasa are present in other transcendental relationships, such as dasya (servitorship), sakhya (fraternity), vatsalya (paternal affection), and the madhura-rasa (conjugal love).

When we speak of non-Krishna, or desire which has no connection with Krishna, this does not mean that anything exists without Krishna. Actually there cannot be anything "non-Krishna" because everything is a product of the energy of Krishna. Since Krishna and His energies are identical, everything is Krishna indirectly.

For example, consciousness is common to every living entity, but when consciousness is purely centred on Krishna (Krishna consciousness), it is pure, and when consciousness is centred on something other than Krishna, or when it is directed to sense gratification, it may be called non-Krishna consciousness. Thus it is in the polluted state that the conception of non-Krishna comes. In the pure state, however, there is nothing but Krishna consciousness.

Active interest in Krishna—the understanding that Krishna is mine or that I am Krishna's, and that therefore my business is to satisfy the senses of Krishna—is typical of a higher stage than the neutrality of the santa-rasa. Simply by understanding the greatness of Krishna, one can achieve the status of santa-rasa, in which the worshipable object may be the impersonal Brahman or Paramatma. Worship of the impersonal Brahman and the Paramatma is conducted by those engaged in empiric philosophical speculation and mystic yoga.

However, when one develops even further in Krishna consciousness, or spiritual understanding, he can appreciate that the Paramatma, the Supersoul, is the eternal worshipable object, and he surrenders unto Him. Bahunam janmanam ante jnanavan mam prapadyate (Bg. 7.19): "After many, many births of worshipping Brahman and Paramatma, when one surrenders unto Vasudeva as the supreme master and accepts himself as the eternal servitor of Vasudeva, he becomes a great transcendentally realized soul." At that time, due to his thick and thin relationship with the Supreme Absolute Truth, one begins to render some sort of transcendental loving service to the Supreme Personality of Godhead. Thus the neutral relationship known as santa-rasa is transformed into dasya-rasa, servitorship.

On the platform of dasya-rasa, the greatest quantity of awe and veneration of the Supreme Lord is exhibited. That is, in the dasya-rasa, the greatness of the Supreme Lord is appreciated. It should be noted here that on the platform of santa-rasa there is no spiritual activity, but on the platform of dasya-rasa, service begins. Thus in the dasya-rasa the quality of the santa-rasa is exhibited, and, in addition, there is consciousness of the transcendental taste of service.

Transcendental qualities are certainly present in the santa-rasa and dasya-rasa, but beyond these there is another quality, confidential attachment, which is pure transcendental love. This loving confidence in the Supreme Personality is technically known as visrambha. On the platform of visrambha, fraternity, there is no sense of awe or veneration towards the Supreme Personality of Godhead. Thus in the transcendental fraternal relationship known as sakhya-rasa, there are three transcendental characteristics: the sense of greatness, the sense of service, and the sense of intimacy without awe or veneration. Thus in the sakhya-rasa, the relationship of fraternity, the transcendental qualities are further increased.

Similarly, on the platform of paternal affection (vatsalya-rasa) there are four qualities. In addition to the three qualities already mentioned, there is the sense that the Supreme Lord is dependent on the mercy of the devotee. As a parent of the Supreme Personality of Godhead, the devotee sometimes chastises the Lord and considers himself to be the Lord's maintainer. This transcendental sense of being the maintainer of the supreme maintainer is very pleasing both to the devotee and to the Supreme Lord.

The Lord instructed Srila Rupa Gosvami to write the transcendental literature named Bhakti-rasamrta-sindhu, the science of devotional service, and indicate therein the substance of these five transcendental relationships. It is explained in that great literature how the transcendental relationship of santa-rasa, taking the shape of unflinching faith in Krishna, is further developed into dasya-rasa with the spirit of service, and then to sakhya-rasa or undeterred fraternity, and further to the transcendental platform of paternal love, wherein one feels himself to be maintaining the Lord. All these relationships culminate on the highest platform of conjugal love (madhura-rasa), wherein all these transcendental relationships exist simultaneously.

Sanatana Gosvami

vande 'nantadbhutaisvaryam
sri-Chaitanya-mahaprabhum
nico 'pi yat-prasadat syad
bhakti-shastra-pravartakah

I offer my respectful obeisances unto Lord Chaitanya Mahaprabhu, by whose mercy even a person in the lowest form of life can find direction in transcendental devotional service to the Lord.

After Lord Chaitanya accepted the renounced order of life (sannyasa), He travelled all over India. During this period He went to Maldah, a district in Bengal. In that area there was a village named Ramakeli, where two government ministers of the Nawab Hussain Shah's regime lived. These two ministers were named Dabira Khasa and Sakara Mallika, and they were later to be renamed Sanatana Gosvami and Rupa Gosvami. Being inspired by Lord Chaitanya, they decided to retire from government service and join His sankirtana movement.

Upon making this decision, the two brothers at once took steps to leave their material engagements, and they appointed two learned brahmanas to perform certain Vedic religious rituals that would enable them to achieve complete freedom for the devotional service of Krishna. These preliminary activities are known as purascarya. These ritualistic functions demand that three times a day one worships and offers respects to his forefathers, offers oblations to a fire, and respectfully offers food to a learned brahmana. Five items—time, worship, offering of respect, offering of oblation into the fire and offering of food to a brahmana—comprise purascarya. This and other rituals are mentioned in the hari-bhakti-vilasa, the authoritative book of directions.

After performing these religious rituals, the younger brother, Sakara Mallika (Rupa Gosvami), returned home with an immense amount of money which he had acquired during his government service. Indeed, the silver and gold coins he brought back filled a large boat. After arriving home, he divided the accumulated wealth first in twain and distributed one part to the brahmanas and Vaishnavas.

Thus for the satisfaction of the Supreme Personality of Godhead, he distributed fifty percent of his accumulated wealth to persons engaged in the Supreme Lord's transcendental loving service. Brahmanas are meant to understand the Absolute Truth, and once they understand the truth and actually engage in the loving service of the Lord, they can be called Vaishnavas. Both brahmanas and Vaishnavas are supposed to be fully engaged in transcendental service, and Rupa Gosvami, considering their important transcendental position, gave them fifty percent of his wealth. The remaining fifty percent was again divided in twain—he distributed one part to his relatives and dependent family members, and the other he kept for personal emergencies.

Such distribution of personal wealth is very instructive for all who desire to be elevated in spiritual knowledge. Generally a person bequeaths all his accumulated wealth to his family members and then retires from family activities in order to make progress in spiritual knowledge. Here, however, we find the behaviour of Rupa Gosvami to be exemplary; he gave fifty percent of his wealth for spiritual purposes. This should serve as an example for everyone. The twenty-five percent of his accumulated wealth which he kept for personal emergencies was deposited with a good business firm, since in those days there were no banks. Ten thousand coins were deposited for expenditures incurred by his elder brother, Sanatana Gosvami.

At this time Rupa Gosvami received information that Lord Chaitanya Mahaprabhu was preparing to proceed to Vrndavana from Jagannath Puri. Rupa Gosvami sent two messengers to get actual information of the Lord's itinerary, and he made his own plans to go to Mathura to meet the Lord. It appears that Rupa Gosvami got permission to join Lord Chaitanya, but

Sanatana Gosvami did not. Therefore Sanatana Gosvami entrusted the responsibilities of his government service to his immediate assistants, and he remained home to study Srimad-Bhagavatam. In fact, he even engaged some ten or twenty learned brahmanas and began an intensive study of Srimad-Bhagavatam in their company. While he was thus engaged, he submitted sick-leave reports to his employer, the Nawab. However, the ruler was so anxious for Sanatana Gosvami's advice in government matters that he suddenly appeared at his house. When the Nawab entered the house where Sanatana Gosvami and the brahmanas were assembled, they all stood up to receive him respectfully, and they offered him a place to sit.

"You have submitted sick reports," the Nawab told Sanatana Gosvami: "But I sent my physician to see you, and he reported that you have no illness at all. Since I did not know why you were submitting sick reports and not attending to your service, I have personally come to see you. Frankly, I am much perturbed by your behaviour. As you know, I completely depend on you and your responsible work in government. I was free to act in other matters because I was depending on you, but if you do not join me, your past devotion will be spoiled. Now, what is your intention? Please tell me."

On hearing this, Sanatana Gosvami replied that he was unable to continue work and that it would be very kind of the Nawab to appoint someone else to execute the work that was entrusted to him. Upon hearing this, the Nawab became very angry and said, "Your elder brother lives like a hunter, and if you also retire from the administration, everything will be finished." It was said that the Nawab used to treat Sanatana Gosvami like a younger brother. Since the Nawab was principally engaged in conquering different parts of the country and also in hunting, he depended largely on Sanatana Gosvami for government administration. Thus he pleaded with him: "If you also retire from government service, how will the administration carry on?"

"You are the governor of Gauda," Sanatana Gosvami replied very gravely, "and you punish different kinds of criminals in different ways. So you are at liberty to punish anyone according to his activity." By this reply Sanatana Gosvami was indicating that since the governor was engaged in hunting animals and in killing men to expand his kingdom, let both of them suffer according to the acts they were performing.

The Nawab was intelligent, and he understood Sanatana Gosvami's purpose. He left the house in an angry mood, and shortly afterward he went off to conquer Orissa. He ordered the arrest of Sanatana Gosvami and commanded that he be held until he returned. Upon learning that his elder brother had been arrested by the Nawab, Rupa Gosvami sent information that ten thousand coins were being held in the custody of a grocer in Gauda (Bengal) and that this money could be used as ransom for his brother.

Sanatana also offered five thousand coins to the keeper of the jail in which he was being held in custody. He advised the jailkeeper to gladly accept the five thousand coins from him and let him go because by accepting the money he would not only be materially benefited but would also be acting very righteously by freeing Sanatana for spiritual purposes.

"Of course I can let you go," the jailkeeper replied, "for you have done many services for me, and you are in government service. However, I'm afraid of the Nawab. What will he do when he hears that you are free? I'll have to explain everything to him. How can I accept such a proposal?" Sanatana then invented a story which the jailkeeper might submit to the Nawab—as to how he had escaped—and he raised his offer to ten thousand coins. Greedy to get the money, the jailkeeper agreed to the proposition and let him go. In the meantime, Rupa Gosvami, with his younger brother Sri Vallabha, had started for Vrndavana to meet Chaitanya Mahaprabhu.

Sanatana then proceeded to go to see the Lord. He did not travel on the open road but went through the jungles until he arrived at a place in Bihar called Patada. There he rested in a hotel, but the hotelkeeper was informed by an astrologer employed there that Sanatana Gosvami had some gold coins with him. The hotelkeeper, desiring to get the money, spoke to Sanatana with seeming respect.

"Just take your rest tonight," he told him, "and in the morning I shall arrange for you to get out of this jungle trap." However, Sanatana was suspicious of his behaviour, and he inquired from his servant Isana whether he had money, and Isana told him that he had seven gold coins. Sanatana did not like the idea of the servant carrying such money. He became angry with him and said, "Why do you carry this death knell on the road?"

Sanatana at once took the gold coins and offered them to the hotelkeeper. He then requested that the hotel-keeper help him through the jungle. He informed him that he was on a special journey for the government and that since he could not travel on the open road, it would be very kind if the hotel-keeper would help him through the jungles and over the mountains.

"I understood that you had eight coins with you, and I was thinking of killing you to take them," the hotel-keeper confessed. "But I can understand that you are such a good man that you don't have to offer me the money."

"If you don't accept these coins, then someone else will take them from me," Sanatana replied. "Someone will kill me for them, so it is better that you take them. I offer them to you." The hotel-keeper then gave him all assistance, and that very night he helped him get past the hills.

When Sanatana emerged from the hills, he requested that his servant go home with the one coin that he still had with him, for Sanatana decided that he would go on alone. After the departure of his servant, Sanatana felt completely free. With torn clothing and a waterpot in hand, he began to proceed toward Lord Chaitanya Mahaprabhu. On the way, he met his rich brother-in-law who was also in the government service and who offered him an excellent blanket, which Sanatana accepted at his special request. Then he parted from him and went on alone to see Chaitanya Mahaprabhu at Benares.

When he reached Benares, he understood that the Lord was there, and he became overjoyed. He was informed by the people that the Lord was staying at the house of

Chandrashekhar Acarya, and Sanatana went there. Although Chaitanya Mahaprabhu was inside the house, He could understand that Sanatana had arrived at the door, and He asked Chandrashekhar to call the man who was sitting there. "He is a Vaishnava, a great devotee of the Lord," Chaitanya Mahaprabhu said. Chandrashekhar came out to see the man, but he saw no Vaishnava at the door. He saw only a man who appeared to be a mendicant.

The Lord then asked to see the mendicant, and when Sanatana entered the courtyard, Lord Chaitanya hurriedly came to see and embrace him. When the Lord embraced him, Sanatana became overwhelmed with spiritual ecstasy, and he said, "My dear Lord, please do not touch me." But both of them embraced each other and began to cry. Seeing Sanatana and Lord Chaitanya acting thus, Chandrashekhar was struck with wonder. At length, Chaitanya Mahaprabhu asked Sanatana to sit down with Him on a bench. He was touching the body of Sanatana with His hand, and Sanatana asked Him again, "My dear Lord, please do not touch me."

"I am touching you just for My purification," the Lord replied, "for you are a great devotee. By your devotional service you can deliver the whole universe and enable everyone to go back to Godhead."

The Lord proceeded to quote a verse from Srimad-Bhagavatam to the effect that a person who is a devotee of Lord Krishna and is one hundred percent engaged in devotional service is far better than a brahmana who is versed in all the Vedic literatures but who does not engage in the devotional service of the Lord. Because he is carrying the Supreme Lord within his heart, the devotee can purify everyplace and everything.

In the Vedic literatures it is also stated that the Supreme Personality of Godhead does not recognize a person who is very learned in all the divisions of the Vedas, but, rather, He likes a person who is a devotee, even though he may be born in a low family. If one offers charity to a brahmana who is not a devotee, the Lord does not accept; but if something is offered to a devotee, the Lord accepts.

In other words, whatever a person wishes to offer the Lord may be given to His devotees. Chaitanya Mahaprabhu also quoted Srimad-Bhagavatam to the effect that if a brahmana is not a devotee of the Supreme Lord, then he is lower than the lowest of the low, even though he may be qualified with the twelve brahminical qualities and born in a high family. A devotee, although born in a candala (dog-eater) family, can purify his whole family for one hundred generations, past and future, by devotional service, whereas a proud brahmana cannot even purify himself. It is said in the Hari-bhakti-sudhodaya (13.2):

aksnoh phalam tvadrsa-darśanam hi
tanoh phalam tvadrsa-gatra-sangah
jihva-phalam tvadrsa-kirtanam hi
sudurlabha bhagavata hi loke

"O devotee of the Lord, to see you is the perfection of the eyes, to touch your body is the perfection of bodily activities, and to glorify your qualities is the perfection of the tongue, for it is very rare to find a pure devotee like you."

The Lord then told Sanatana that Krishna is very merciful and is the deliverer of fallen souls. "He has saved you from Maharaurava," the Lord said. This Maharaurava, or hell, is described in Srimad-Bhagavatam as a place meant for persons who are engaged in killing animals, for it is stated there that butchers or animal eaters go to that hell.

"I do not know the mercy of Krishna," Sanatana replied, "but I can understand that Your mercy upon me is causeless. You have delivered me from the entanglement of material life."

Then the Lord inquired: "How did you get free from your custody? I understand that you were arrested." Sanatana then narrated the whole story of his release. "I have seen your two brothers," the Lord then informed him, "and I have advised them to proceed toward Vrndavana."

Lord Chaitanya then introduced Chandrashekhar and Tapana Mishra to Sanatana, and Tapana Mishra pleasantly invited Sanatana to dine with him. The Lord requested Chandrashekhar to take Sanatana to a barber and make him "gentle," for Sanatana had grown a long beard which Sri Chaitanya Mahaprabhu did not like. He not only asked Chandrashekhar to provide Sanatana with a bath and clean shave but with a change of clothes as well.

After bathing, Chandrashekhar gave him some good cloth. When Lord Chaitanya was informed that Sanatana did not accept new garments but later accepted only some used garments from Tapana Mishra, He was very glad. The Lord went to Tapana Mishra's house for lunch and asked him to keep food for Sanatana. Tapana Mishra did not offer Sanatana food immediately, however, but after the Lord finished His eating there were some remnants of His foodstuff, and that was offered to Sanatana while the Lord took His rest.

After resting, Lord Chaitanya introduced one Maharastriya brahmana, a devotee of His, to Sanatana, and that Maharastriya brahmana invited Sanatana to accept lunch daily at his place as long as he remained in Benares.

"As long as I remain in Benares, I will beg from door to door," Sanatana said. "But the Lord will be so good as to accept this invitation for daily lunch at your house."

Lord Chaitanya was very much pleased by this behaviour of Sanatana, but He noticed the valuable blanket that was given to him by his brother-in-law while en route to Benares. Although Lord Chaitanya was overlooking the blanket, Sanatana understood that He did not approve of such a valuable garment on his body, and therefore Sanatana decided to get rid of it. He immediately went to the bank of the Ganges, and there he saw a mendicant washing an old quilt. When Sanatana asked him to trade the old quilt for the valuable blanket, the poor mendicant thought that Sanatana was joking with him. "How is this?" the mendicant upbraided him. "You appear to be a very nice gentleman, but you are mocking me in this unmannerly way.

"I am not joking with you," Sanatana informed him. "I am very serious. Will you kindly exchange that torn quilt for this blanket?" Finally the mendicant exchanged his torn quilt for the blanket, and Sanatana returned to the Lord.

"Where is your valuable blanket?" the Lord immediately inquired. Sanatana informed Him about the exchange, and the Lord loved him for this and thanked him. "You are intelligent enough, and you have now exhausted all your attraction for material wealth."

In other words, the Lord accepts a person for devotional service only when he is completely free from all materialistic possessions. The Lord then told Sanatana: "It would not look good for you to be a mendicant and beg from door to door with such a valuable blanket on your body. It is contradictory and people would look on it with abhorrence."

"Whatever I am doing to become free from material attachment is all Your mercy," Sanatana replied. The Lord was very much pleased with him, and both of them discussed spiritual advancement.

Previous to this meeting between Lord Chaitanya and Sanatana Gosvami, the Lord met a householder devotee named Ramananda Raya. At that meeting, which is discussed in a later chapter, Lord Chaitanya asked Ramananda Raya questions, and Ramananda replied as if he were the teacher of the Lord. However, in this case Sanatana put questions to the Lord, and the Lord answered them Himself.

The instructions and teachings of Lord Chaitanya are very important for people in general. He teaches the process of devotional service, which is the constitutional occupation of every living entity, for it is every man's duty to advance in spiritual science. Many subjects were thoroughly discussed in the talks between Lord Chaitanya and Sanatana Gosvami. Due to the mercy of Lord Chaitanya, Sanatana was able to put important questions before Him, and these questions were replied to properly.

By the meeting of Sanatana and Lord Chaitanya, we learn that in order to understand spiritual subject matters, one must approach a spiritual master like Lord Chaitanya Mahaprabhu and make submissive inquiries. It is also confirmed in Bhagwad Gita (Bg. 4.34) that one should approach a man of authority and learn the science of spiritual life from him.

Teachings to Sanatana Gosvami

From the instructions of Lord Chaitanya to Sanatana Gosvami we can understand the science of God as it relates to God's transcendental form, His opulences, and His devotional service. Indeed, everything is being explained to Sanatana Gosvami by the Lord Himself. At that time, Sanatana fell at the feet of the Lord and with great humility asked about his own real identity. "I am born of a lower family," Sanatana said. "My associations are all abominable, and I am fallen, the most wretched of mankind. I was suffering in the dark well of material enjoyment, and I never knew the actual goal of my life. Indeed, I do not even know what is beneficial for me. Although I am what is known in the world as a great learned man, I am in fact so much of a fool that I myself even think that I am learned. You have accepted me as Your servant, and You have delivered me from the entanglement of material life. Now please tell me what my duty is in this liberated state."

By this plea, we can understand that liberation is not the final word in perfection. There must be activities in liberation. Sanatana clearly says, "You have saved me from the material existence. Now, after liberation, what is my duty?" Sanatana further inquired, "Who am I? Why are the threefold miseries always giving me trouble? And finally, tell me how I can be relieved from this material entanglement? I do not know how to question You about the advancement of spiritual life, but I beg that You kindly, mercifully, let me know everything that I need know."

This is the process of accepting a spiritual master. One should approach a spiritual master, humbly submit to him and then inquire from him about one's spiritual progress. The Lord was pleased by Sanatana's submissive behaviour, and He replied, "You have already received benediction from Lord Krishna, and therefore you know everything and are free from all the miseries of material existence." The Lord further pointed out that because Sanatana was in Krishna consciousness, he was naturally, by the grace of Krishna, already conversant with everything. "Because you are a humble devotee," the Lord continued, "you are asking Me to confirm what you already know. This is very nice." These are the characteristics of a true devotee. In the Narada-bhakti-sutra it is said that one who is very serious about developing Krishna consciousness has his desire to understand Krishna fulfilled very soon by the grace of the Lord.

"You are a suitable person to protect the devotional service of the Lord," Chaitanya Mahaprabhu continued. "Therefore it is My duty to instruct you in the science of God, and I will explain everything to you step by step."

It is the duty of a disciple approaching a spiritual master to inquire about his constitutional position. In conformity to that spiritual process, Sanatana has already asked, "What am l, and why am I suffering from the threefold miseries?" The threefold miseries are called adhyatmika, adhibhautika, and adhidaivika. The word adhyatmika refers to those miseries caused by the mind and body. Sometimes the living entity suffers bodily, and sometimes he is distressed mentally. Both are adhyatmika miseries. We experience these miseries even in the womb of our mother.

As we well know, there are many types of miseries that take advantage of the delicate human body and give us pain. Miseries inflicted by other living entities are called adhibhautika. These living entities need not even be large, for there are many—such as bugs—that can make us miserable even while we are sleeping in bed. There are many insignificant living entities, like cockroaches, that sometimes give us pain, and there are also other living entities who are born on different kinds of planets and who give us miseries. As far as the adhidaivika miseries are concerned, these are natural disasters that originate with the demigods of the higher planets. For instance, we sometimes suffer from severe cold or hot weather, from a thunderbolt, or from earthquakes, tornadoes, droughts and many natural disasters. In any case, we are always suffering from either one or a combination of these three kinds of miseries.

Sanatana's inquiry was therefore an intelligent one. "What is the position of the living entities?" he asked. "Why are they always undergoing these three kinds of miseries?" Sanatana had admitted his weakness. Although he was known by the masses of people as a greatly learned man (and actually he was a highly learned Sanskrit scholar), and although he accepted this designation, he did not actually know what his constitutional position really was and just why he was subjected to the threefold miseries.

Approaching a spiritual master is not just a fashion but is a necessity for one who is seriously conscious of the material miseries and who wants to be free of them. It is the duty of such a person to approach a spiritual master. In this regard, we should note similar circumstances in Bhagwad Gita. When Arjuna was perplexed by so many problems involving whether to fight or not, he accepted Lord Krishna as his spiritual master. It was also a case of the supreme spiritual master instructing Arjuna about the constitutional position of the living entity.

In Bhagwad Gita we are informed that the constitutional nature of the individual entity is spirit soul. He is not matter. As spirit soul, he is part and parcel of the supreme soul, the Absolute Truth, the Personality of Godhead. We also learn that it is the duty of the spirit soul to surrender, for only then can he be happy. The last instruction of Bhagwad Gita is that the spirit soul surrender completely unto the supreme soul, Krishna, and in that way realize happiness.

Here also, Lord Chaitanya, answering the questions of Sanatana, repeats the same truth. There is a difference, however. Here Lord Chaitanya does not give the information about the spirit soul that is already described in Bhagwad Gita. Rather, He begins from the point where Krishna ended His instruction. It is accepted by great devotees that Lord Chaitanya is Krishna Himself, and from this point of view He begins His instruction to Sanatana from the point where He ended His instructions to Arjuna in Bhagwad Gita.

"Your constitutional position is that you are pure living soul," the Lord told Sanatana. "This material body cannot be identified with your real self; nor is your mind your real identity, nor your intelligence, nor false ego. Your identity is that of eternal servitor of the Supreme Lord Krishna. Your position is that you're transcendental. The superior energy of Krishna is spiritual in constitution, and the inferior external energy is material. Since you are between the material energy and the spiritual energy, your position is marginal. Belonging to the marginal potency of Krishna, you are simultaneously one with and different from Krishna. Because you are spirit, you are not different from Krishna, and because you are only a minute particle of Krishna, you are different from Him."

This simultaneous oneness and difference always exists in the relationship between the living entities and the Supreme Lord. From the marginal position of the living entities, this conception of "simultaneously one and different" can be understood. The living entity is just like a molecular particle of sunshine, whereas Krishna may be compared to the blazing, shining sun itself. Lord Chaitanya compared the living entities to blazing sparks from a fire and the Supreme Lord to the blazing fire of the sun. In this connection, the Lord cites a verse from Vishnu purana (1.22.52):

eka-desa-sthitasyagner
jyotsna vistarini yatha
parasya brahmanah shaktis
tathedam akhilam jagat

"Everything that is manifested within this cosmic world is but the energy of the Supreme Lord. As fire emanating from one place diffuses its illumination and heat all around, so the Lord, although situated in one place in the spiritual world, manifests His different energies everywhere. Indeed, the whole cosmic creation is composed of different manifestations of His energy."

The energy of the Supreme Lord is transcendental and spiritual, and the living entities are part and parcel of that energy. There is another energy, however, called material energy, which is covered by the cloud of ignorance. This energy, which is material nature, is divided into three modes, or gunas (goodness, passion and ignorance). Lord Chaitanya quoted from Vishnu purana (1.3.2) to the effect that all inconceivable energies reside in the Supreme Personality of the Lord and that the whole cosmic manifestation acts due to the Lord's inconceivable energy. The Lord also said that the living entities are known as ksetrajna, or "knowers of the field of activities." In the Thirteenth Chapter of Bhagwad Gita, the body is described as the field of activities, and the living entity as ksetrajna, the knower of that field. Although the living entity is constitutionally conversant with spiritual energy, or has the potency to understand spiritual energy, he is covered by the material energy and consequently identifies the body with the self. This false identification is called "false ego." Deluded by this false ego, the bewildered living entity in material existence changes his different bodies and suffers various kinds of miseries. Knowledge of the living entity's true position is possessed to different extents by different types of living entities.

In other words, it is to be understood that the living entity is part and parcel of the spiritual energy of the Supreme Lord. Because the material energy is inferior, man has the ability to get uncovered from this material energy and utilize the spiritual energy. It is stated in Bhagwad Gita that the superior energy is covered by the inferior energy. Due to this covering, the living entity is subjected to the miseries of the material world, and, in proportion to the different degrees of passion and ignorance, he suffers material miseries. Those who are a little enlightened suffer less, but on the whole everyone is subjected to material miseries due to being covered by the material energy.

Chaitanya Mahaprabhu also quoted from the Seventh Chapter of Bhagwad Gita in which it is stated that earth, water, fire, air, ether, mind, intelligence and ego all combine together to form the inferior energy of the Supreme Lord. The superior energy, however, is the real identity of the living being, and it is because of that energy that the whole material world functions. The cosmic manifestation, which is made of material elements, has no power to act unless it is moved by the superior energy, the living entity. It can actually be said that the conditioned life of the living entity is due to forgetfulness of his relationship with the Supreme Lord in the superior energy. When that relationship is forgotten, conditional life is the result. Only when man revives his real identity, that of eternal servitor to the Lord, does he become liberated.

The Wise Man

Since no one can trace the history of the living entity's entanglement in material energy, the Lord says that it is beginningless. By beginningless it is meant that conditional life exists prior to the creation; it is simply manifested during and after the creation. Due to forgetfulness of his nature, the living entity, although spirit, suffers all kinds of miseries in material existence. It should be understood that there are also living entities who are not entangled in this material energy but are situated in the spiritual world. They are called liberated souls and are always engaged in Krishna consciousness, devotional service.

The activities of those who are conditioned by material nature are taken into account, and in their next life, according to these activities, they are offered different types of material bodies. In the material world the conditioned spirit soul is subjected to various rewards and punishments. When he is rewarded for his righteous activities, he is elevated to the higher planets where he becomes one of the many demigods, and when he is punished for his abominable activities, he is thrown into hellish planets where he suffers the miseries of material existence more acutely. Chaitanya Mahaprabhu gives a very nice example of this punishment.

Formerly a king used to punish a criminal by dunking him in the river, raising him up again for breath and then again dunking him in the water. Material nature punishes and rewards the individual entity in just the same way. When he is punished, he is dunked in the water of material miseries, and when he is rewarded, he is taken out of it for some time. Elevation to the higher planets or to a higher life status is never permanent. One must again come down to be submerged in the water. All this is constantly going on in this material existence; sometimes one is elevated to higher planetary systems, and sometimes one is thrown into the hellish condition of material life.

In this regard Chaitanya Mahaprabhu recites a verse from Srimad-Bhagavatam taken from the instructions of Narada Muni to Vasudeva, the father of Krishna (Bhag. 11.2.37):

bhayam dvitiyabhinivesatah syad
isad apetasya viparyayo 'smrtih
tan-mayayato budha abhajet tam
bhaktyaikayesam guru-devatatma

In this quotation from the nine sages who were instructing Maharaja Nimi, maya is defined as "forgetfulness of one's relationship with Krishna." Actually, maya means "that which is not." It has no existence. Thus it is false to think that the living entity has no connection with the Supreme Lord. He may not believe in the existence of God, or he may think that he has no relationship with God, but these are all "illusions," or maya. Due to absorption in this false conception of life, man is always fearful and full of anxieties.

In other words, a godless concept of life is maya. One who is actually learned in the Vedic literatures surrenders unto the Supreme Lord with great devotion and accepts Him as the supreme goal. When a living entity forgets the constitutional nature of his relationship with God, he is at once overwhelmed by the external energy. This is the cause of his false ego, his false identification of the body with the self. Indeed, his whole conception of the material universe arises from this false identification with the body, for he becomes attached to the body and its by-products. To escape this entanglement, he has only to perform his duty and to surrender unto the Supreme Lord with intelligence and devotion and with sincere Krishna consciousness.

A conditioned soul falsely thinks himself happy in the material world, but if he is favoured by the instructions of an unalloyed devotee, he gives up his desire for material enjoyment and becomes enlightened in Krishna consciousness. As soon as one enters into Krishna consciousness, his desire for material enjoyment is at once vanquished, and he gradually becomes free from material entanglement. There is no question of darkness where there is light, and Krishna consciousness is the light that dispels the darkness of material sense enjoyment.

A Krishna conscious person is never under the false conception that he is one with God. Knowing that he would not be happy by working for himself, he engages all his energies in the service of the Supreme Lord and thereby gains release from the clutches of illusory material energy. In this connection, Chaitanya Mahaprabhu quotes the following verse from Bhagwad Gita:

daivi hy esa guna-mayi
mama maya duratyaya
mam eva ye prapadyante
mayam etam taranti te.

"The divine energy of Mine, consisting of the three modes of material nature, is difficult to overcome. But those who have surrendered unto Me can easily cross beyond it." (Bg. 7.14)

Chaitanya Mahaprabhu went on to teach that for each and every moment he is engaged in some fruitive activity, the conditioned soul forgets his real identity. Sometimes when he is fatigued, when he is tired of material activities, he wants liberation and hankers to become one with the Supreme Lord, but at other times he thinks that by working hard to gratify his senses he will be happy. In either case, he is covered by material energy. For the enlightenment of such bewildered conditioned souls, the Supreme Lord has presented voluminous Vedic literatures such as the Vedas, the puranas and the Vedanta-sutra. These are all intended to guide the human being back to Godhead. Chaitanya Mahaprabhu has given further instructions by explaining that when a conditioned soul is accepted by the mercy of the spiritual master and is guided by the Supersoul and the various Vedic scriptures, he becomes enlightened and makes progress in spiritual realization. It is because Lord Krishna is always merciful upon His devotees that He has presented all these Vedic literatures by which one can understand his relationship with Him and can act on the basis of that relationship. In this way one is gifted with the ultimate goal of life.

Actually every living entity is destined to reach the Supreme Lord. Indeed, it is possible for everyone to understand his relationship with the Supreme. The execution of duties to attain perfection is known as devotional service, and in maturity such devotional service becomes love of God, the factual goal of life for every living being. Actually the living entity is not intended to achieve success in religious rituals, economic development or sense enjoyment. The living entity should not even desire success in liberation, what to speak of success in religion, economics and sense enjoyment. One's real desire should only be to achieve the stage of loving transcendental service to the Lord. The all-attractive features of Lord Krishna help one in attaining this transcendental service, and it is by such service in Krishna consciousness that one can realize the relationship between Krishna and himself.

Concerning man's search for the ultimate goal of life, Chaitanya Mahaprabhu relates a story from the commentary of Madhva which occurs in the Fifth Canto of Srimad-Bhagavatam (Madhva-bhasya) Sarvajna to a poor man who came to him to have his future told. When Sarvajna saw the horoscope of the man, he was at once astonished that the man was so poor, and he said to him, "Why are you so unhappy? From your horoscope I can see that you have a hidden treasure left to you by your father. However, the horoscope indicates that your father could not disclose this to you because he died in a foreign place, but now you can search out this treasure and be happy." This story is cited because the living entity is suffering due to his ignorance of the hidden treasure of his Supreme Father, Krishna. That treasure is love of Godhead, and in every Vedic scripture the conditioned soul is advised to find it. As stated in Bhagwad Gita, although the conditioned soul is the son of the wealthiest personality—the Personality of Godhead—he does not realize it. Therefore Vedic literatures are given to him to help him search out his father and his paternal property.

The astrologer Sarvajna further advised the poor man: "Don't dig on the southern side of your house to find the treasure, for if you do so you will be attacked by a poisonous wasp and will be baffled. The search should be conducted on the eastern side where there is actual light, which is called devotional service or Krishna consciousness. On the southern side there are Vedic rituals, and on the western side there is mental speculation, and on the northern side there is meditational yoga."

Sarvajna's advice should be carefully noted by everyone. If one searches for the ultimate goal by the ritualistic process, he will surely be baffled. Such a process involves the performance of rituals under the guidance of a priest who takes money in exchange for service. A man may think he will be happy by performing such rituals, but actually if he does gain some result from them, it is only temporary. His material distresses will continue. Thus he will never become truly happy by following the ritualistic process. Instead, he will simply increase his material pangs more and more. The same may be said for digging on the northern side, or searching for the treasure by means of the meditational yoga process. By this process a person thinks of becoming one with the Supreme Lord, but this merging into the Supreme is like being swallowed by a large serpent. Sometimes a large serpent swallows a smaller one, and merging into the spiritual existence of the Supreme is analogous. While the small serpent is searching after perfection, he is swallowed. Obviously there is no solution here.

On the western side there is also an impediment in the form of a yaksa, an evil spirit who protects the treasure. The idea is that a hidden treasure can never be found by one who asks the favour of a yaksa in order to attain it. The result is that one will simply be killed. This yaksa is the speculative mind, and in this case the speculative process of self-realization, or the jnana process, is also suicidal. The only possibility then is to search for the hidden treasure on the eastern side by the process of devotional service in full Krishna consciousness. Indeed, that process of devotional service is the perpetual hidden treasure, and when one attains to it, he becomes perpetually rich. One who is poor in devotional service to Krishna is always in need of material gain. Sometimes he suffers the bites of poisonous creatures, and sometimes he is baffled; sometimes he follows the philosophy of monism and thereby loses his identity, and sometimes he is swallowed by a large serpent. It is only by abandoning all this and becoming fixed in Krishna consciousness, devotional service to the Lord, that one can actually achieve the perfection of life.

2

Lord Chaitanya, the Original Personality of Godhead

Following the footsteps of Kaviraja Krishnadasa Gosvami, we offer our respectful obeisances unto the lotus feet of Lord Chaitanya.

Lord Chaitanya is described as follows: He is the only shelter for the forlorn, for the most fallen, and He is the only hope for those who are completely devoid of spiritual knowledge. Let us try to discuss His great contribution of devotional service.

The supremely powerful Lord Krishna is manifest in five different potencies. Although He is one without a second, in order to serve five specific spiritual purposes, He is manifest in five ways. Such diversity is eternal and blissful, in contrast to the conception of monotonous oneness. From Vedic literatures we can understand that the Absolute Truth, the Supreme Personality of Godhead, eternally exists with His diverse energies. Lord Chaitanya appeared with full diverse energies, and they are five in number; therefore Lord Chaitanya is said to be Krishna with diverse energies.

There is no difference between the energy and the energetic in regard to the Lord's appearance as Sri Chaitanya Mahaprabhu and His four associates—Nityananda Prabhu, Advaita Prabhu, Gadadhara, and Srivasa. Amongst these five diverse manifestations of the Supreme Lord (as the Lord Himself and His incarnation, expansion and energies) there is no spiritual difference. They are five in one Absolute Truth. For the sake of relishing transcendental flavours in the Absolute Truth, there are five diverse manifestations. These are called the form of a devotee, the identity of a devotee, the incarnation of a devotee, the pure devotee and devotional energy.

Out of the five diversities in the Absolute Truth, the form of Lord Chaitanya is that of the original Personality of Godhead, Krishna. Lord Nityananda is the manifestation of the first expansion of the Supreme Lord. Similarly, Advaita Prabhu is an incarnation of the Supreme Lord. These three—Chaitanya, Nityananda and Advaita—belong to the category of Vishnu-tattva, or the Supreme Absolute Truth. Srivasa represents the pure devotee, and Gadadhara represents the internal energy of the Lord for the advancement of pure devotion.

Therefore Gadadhara and Srivasa, although included in Vishnu-tattva, are dependent, diverse energies of the Supreme Lord. In other words, they are not different from the energetic, but they are manifest diversely for the sake of relishing transcendental relationships. The whole process of devotional service involves a transcendental reciprocation in the relationship between the worshipper and the worshipped. Without such a diverse exchange of transcendental flavours, devotional service has no meaning.

In the Vedic literature (Katha Upanishad) it is stated that the Supreme Lord is the supreme living entity amongst all living entities. There are innumerable living entities, but there is one living entity who is the Supreme Absolute Godhead. The difference between the singular living entity and the plural living entities is that the singular living entity is the Lord of all. Lord Chaitanya is that supreme living entity, and He descended to reclaim the innumerable fallen living entities.

In other words, the specific purpose of Lord Chaitanya's advent was to establish the Vedic fact that there is one Supreme Personality of Godhead predominating over and maintaining the innumerable personalities of all living entities. Because the impersonalist (Mayavadi) philosophers cannot understand this, Lord Chaitanya advented Himself to enlighten the people in general about the real nature of the relationship between the Supreme and the many entities.

In Bhagwad Gita Krishna's last instructions were that everyone should give up all other engagements and render devotional service unto Him. However, after Krishna's disappearance, less intelligent people misunderstood Him. They became contaminated with the Mayavadi philosophy, which produced so many mental speculators that people forgot the actual position of the Absolute Truth and the living entity. Therefore Lord Sri Krishna Himself, as Lord Chaitanya, again appeared to teach the fallen souls of this material world the way to approach Lord Krishna.

The Bhagwad Gita teaches that one should give up everything and be done with this world of material attachment. A pure devotee of Lord Krishna and one who follows the philosophy of Lord Chaitanya are one and the same. Chaitanya's philosophy is that one should give up everything and worship God, Krishna. Krishna, as the Supreme Lord, the Personality of Godhead, spoke the same words, indicating Himself as the Supreme Lord. However, the Mayavadi philosophers misunderstood Him. Therefore Lord Chaitanya, to clarify the situation, reiterated Lord Krishna's message: One should not declare himself to be as good as Krishna but should worship Krishna as the Supreme Lord.

We make a great mistake if we accept Lord Chaitanya as a conditioned soul. He is to be understood as the Supreme Absolute Truth, the Personality of Godhead, Sri Krishna Himself. It is therefore said of Lord Chaitanya in Chaitanya-caritamrta: "Krishna is now present in His five diverse manifestations." Unless one is situated in uncontaminated goodness, it is very difficult to understand Lord Chaitanya as the Supreme Personality of Godhead Himself. Thus in order to understand Lord Chaitanya, one has to follow the direct disciples of Lord Chaitanya—the six Gosvamis—and especially the path chalked out by Srila Jiva Gosvami.

The most astonishing fact is that Lord Chaitanya, although the Supreme Personality of Godhead, Krishna, never displayed Himself as Krishna. Rather, whenever He was detected by intelligent devotees as Lord Krishna and was addressed as Lord Krishna, He denied it. Indeed, He sometimes placed His hands over His ears, protesting that one should not be addressed as the Supreme Lord. Indirectly, He was teaching the Mayavadi philosophers that one should not falsely pose himself as the Supreme Lord and thereby misguide people. Nor should followers be foolish enough to accept anyone and everyone as the Supreme Personality of Godhead.

One should test by consulting scriptures and by seeing the activities of the person in question. One should not, however, mistake Lord Chaitanya and His five diverse manifestations as ordinary human beings. Lord Chaitanya is the Supreme Personality of Godhead, Krishna Himself. The beauty of Lord Chaitanya is that although He is the Supreme Lord, He came as a great devotee to teach all conditioned souls how devotional service should be rendered. Conditioned souls who are interested in devotional service should follow in the exemplary footsteps of Lord Chaitanya in order to learn how Krishna can be achieved by devotional service. Thus the Supreme Lord Himself teaches the conditioned soul how He should be approached in devotional service.

By analytically studying the five diverse manifestations of the Supreme Lord, we can come to know that Lord Sri Chaitanya Mahaprabhu is the Supreme Absolute and that Lord Nityananda is an immediate expansion of the Supreme Absolute Truth. We can also come to understand that Advaita Prabhu is also in the category of the Supreme Personality of Godhead but is subordinate to Lord Chaitanya and Nityananda Prabhu. The Supreme Personality of Godhead and His immediate subordinate expansions are worshipable by the other two—namely the representation of the internal potency and the representation of the marginal potency. The representation of the internal potency, Gadadhara, represents the confidential devotee, and the representation of the marginal potency is the pure devotee. Both of these are worshippers of the other three categories, but all of them are engaged in the transcendental service of the Supreme Personality of Godhead, Sri Chaitanya Mahaprabhu.

There is a specific difference between the pure devotee and the confidential devotee. Different potencies of the Lord are engaged in serving the Supreme Lord in different transcendental relationships. They are situated in conjugal love, in parental affection, in friendship and in servitude. By impartially judging, one can find that the internal potencies of the Supreme Lord who are engaged in conjugal love with the Lord are the best of all devotees.

Thus both internal devotees and confidential devotees are attracted by the conjugal love of the Supreme Absolute Truth. These are the most confidential devotees of Lord Chaitanya. Other pure devotees, who are more or less attached to Sri Nityananda Prabhu and Advaita Prabhu, are attracted by other transcendental relationships, such as parental affection, friendship and servitorship. When such devotees are attached to the activities of Lord Chaitanya, they at once become confidential devotees in conjugal love with the Supreme Lord.

There is a very nice song sung by Sri Narottama dasa Thakura, a great devotee and acharya in the disciplic succession from Lord Chaitanya. Narottama dasa sings: "When will there be transcendental vibrations all over my body simply by my hearing the name of Gauranga? When will tears incessantly flow from my eyes simply by my uttering the names of the Lord? When will Lord Nityananda have mercy upon me, and when will all my desires for material enjoyment become insignificant? When shall I be purified by giving up all contaminations of material enjoyment? And when shall I be able to see the transcendental abode, Vrndavana? When shall I be eager to accept the six Gosvamis as my prime guidance? And when will I be able to understand the conjugal love of Krishna?" No one should be eager to understand the conjugal love of Krishna without undergoing disciplinary training under the six Gosvamis of Vrndavana.

The sankirtana movement inaugurated by Lord Chaitanya is a transcendental pastime of the Lord. "By it I live simultaneously to preach and popularize this movement in the material world." In that sankirtana movement of Lord Chaitanya, Nityananda and Advaita are His expansions, and Gadadhara and Srivasa are His internal and marginal potencies. The living entities are also called marginal potency because they have, potentially, two attitudes—namely the tendency to surrender unto Krishna and the tendency to become independent of Him. Due to the propensity for material enjoyment, the living entity becomes contaminated by the material world. When a living entity is dominated by a desire for material enjoyment and becomes entangled in material life, he is subjected to the threefold miseries of material existence. He is just like a seed sown in the earth. If a seed is overpowered by too much water, there is no possibility of its fructifying. Similarly, if a man is captivated by material enjoyment, and even if the seed of such enjoyment is within the heart of the conditioned soul, he can be overpowered by a flood of transcendental activities performed in love of God. In this way his potential seed cannot fructify into a conditional life of material existence. The conditioned living entities in the material world, especially in the present age of Kali, are overpowered by the flood of love of God inaugurated by Lord Chaitanya and His associates.

In this connection there is a verse written by His Holiness Prabodhananda Saraswati in his book Sri Chaitanya-candramrta which states that materialistic persons are very enthusiastic to maintain their family members, wife and children, and that there are also many mystic speculators who are engaged in speculating about liberation from the miseries of material life and who therefore undergo various austerities and penances. But those who have discovered the greatest transcendental flavour in the movement of Lord Chaitanya Mahaprabhu no longer have a taste for such activities.

Those who are under the impression that there is material contamination in the form of the Supreme Personality of Godhead and in His devotional service are called Mayavadi. According to their imperfect speculation, the impersonal Brahman is considered to be the only existence in the cosmic manifestation. As soon as the Supreme Personality of Godhead is introduced, they consider that His personality arises from maya, or the external material energy. Such persons consider all incarnations of the Supreme Lord to be contaminated by this material nature. According to them, the material body and the activities of matter which identify the living entity are all material manifestations. According to them, liberation means the end of individual identification, or of the pure living entity. In other words, the Mayavadis maintain that when a living entity is liberated, he becomes one with the supreme impersonal Brahman.

According to such Mayavadi philosophy, the Personality of Godhead, His abode, His devotional service and His emotional devotees are all under the spell of maya and are consequently subjected to the material condition. Those who forget the transcendental nature of the Supreme Lord, His abode, His devotional service and His devotees consider all these to be but manifestations of material activity. When one thinks that there is a possibility of arguing about transcendence, he is called an agnostic, and when he thinks that there is a possibility to criticize transcendence, he is called an atheist. Lord Chaitanya wanted to accept all kinds of agnostics, atheists, skeptics and unfaithfuls and swallow them in the flood of love of God. Therefore He accepted the renounced order of life in order to attract all these forces.

Lord Chaitanya remained a householder until His twenty-fourth year, and in the twenty-fifth year of His life, He accepted the renounced order. After accepting the renounced order (sannyasa), He attracted many other sannyasis. When He had been spreading the sankirtana movement as a family man, many Mayavadi sannyasis did not take His movement very seriously, but after the Lord accepted the sannyasa order of life, He delivered speculative students, atheists and those who are attached to fruitive activities and unnecessary criticism. The Lord was so kind that He accepted all these people and delivered to them the most important factor in life: love of God.

To fulfil His mission of bestowing love of God upon conditioned souls, Lord Chaitanya devised many methods to attract those people disinterested in love of God. After He accepted the renounced order, all agnostics, critics, atheists and mental speculators became His students and followers. Even many who were not Hindus and who did not follow the Vedic principles accepted Lord Chaitanya as the supreme teacher. The only persons who avoided the mercy of Sri Chaitanya Mahaprabhu were those sannyasis who were known as the Mayavadi philosophers of Benares. The plight of such Mayavadi philosophers is described by Sri Bhaktisiddhanta Saraswati Gosvami: "The Mayavadi philosophers of Benares were less intelligent because they wanted to measure everything by direct perception. But everything that is perceived is calculated by material perception. The Absolute Truth is transcendence, but according to them there is no variegatedness in transcendence, for they say that anything that is full of variegatedness is maya."

During Chaitanya Mahaprabhu's time there were also other impersonalist philosophers known as the Mayavadi philosophers of Sarnath. Sarnath is a place near Benares where Buddhist philosophers used to reside, and even today many stupas of the Buddhist Mayavadis can be seen. The Mayavadi philosophers of Sarnath are different from the impersonalists who believe in the impersonal manifestation of Brahman. According to the Sarnath philosophers, there is no spiritual existence at all. The fact is that both the Mayavadi philosophers of Benares and the philosophers of Sarnath are entrapped by material nature. None of them actually know the nature of Absolute Transcendence. Although superficially accepting the Vedic principles and considering themselves to be transcendentalists, the philosophers of Benares do not accept spiritual variegatedness. Because they have no information about devotional service, they are called nondevotees, or those who are against the devotional service of Lord Krishna.

The impersonalists speculate on the Supreme Personality of Godhead and His devotees and subject them to the tests of direct perception. However, the Lord, His devotee and His devotional service are not subject to direct perception. In other words, spiritual variegatedness is unknown to the Mayavadi philosophy; therefore all the Mayavadi philosophers and sannyasis criticized Lord Chaitanya when He was conducting His sankirtana movement.

They were surprised to see Lord Chaitanya after He accepted His sannyasa order from Keshava Bharati, for Keshava Bharati belonged to the Mayavadi school. Since Lord Chaitanya therefore belonged to the Mayavadi sect of sannyasis, the Mayavadis were surprised to see Him engaged in chanting and dancing instead of hearing or reading Vedanta, as is the custom. The Mayavadi philosophers are very fond of Vedanta, and they misinterpret it in their own way. Instead of understanding their own position, they criticized Lord Chaitanya as an unauthorized sanyasi, arguing that because He was a sentimentalist, He was not actually a bona fide sanyasi.

All these criticisms were carried to Lord Chaitanya when He was at Benares, and He was not at all surprised at them. He even smiled when the news was carried to Him. He did not associate with the Mayavadi sannyasis but remained alone and executed His own mission. After staying for some days in Benares, He started for Mathura.

The Conversations with Prakasananda

According to the principles of the Mayavadi sannyasis, singing, dancing and playing musical instruments are strictly prohibited, for they are considered to be sinful activities. The Mayavadi sanyasi is simply supposed to engage in the study of Vedanta. Therefore when the Mayavadi sannyasis in Benares saw that Lord Chaitanya was indulging in singing, dancing, playing musical instruments and always chanting Hare Krishna, Hare Krishna, Krishna Krishna, Hare Hare. Hare Rama, Hare Rama, Rama Rama, Hare Hare, they concluded that He was not educated and that, out of sentiment, He was misleading His followers. Shankaracharya's injunction was that a sanyasi should always be engaged in the study of Vedanta and that he should be satisfied by simply having one cloth and nothing more. Because Lord Chaitanya neither studied Vedanta formally nor ceased from singing and dancing He was criticized by all the sannyasis at Benares, as well as by their householder followers.

When Lord Chaitanya received news of this criticism from His students and disciples, He simply smiled and started for Mathura and Vrndavana. When He returned again to Benares on His way from Mathura to Jagannath Puri, He stayed at the house of Chandrashekhar, who was considered a sudra because he was a clerk. In spite of this, Lord Chaitanya Mahaprabhu made His residence at his home. Lord Chaitanya made no distinctions between brahmanas and sudras; He accepted anyone who was devoted. Customarily, a sanyasi is supposed to take shelter and eat in the home of a brahmana, but Chaitanya Mahaprabhu, as the independent Supreme Personality of Godhead, used His own discretion and decided to stay at Chandrashekhar's house.

In those days, by misusing their brahminical heritage, the brahmanas passed a law to the effect that anyone not born in a brahmana family was to be considered a sudra. Thus even the ksatriyas and vaidyas were also considered sudras. Because the vaidyas were supposed to be descendants of brahmana fathers and sudra wives, they were sometimes called sudras. Thus Chandrashekhar Acharya, although born in a vaidya family, was called a sudra in Benares. As long as Lord Chaitanya stayed in Benares, He remained at Chandrashekhar's home, and He took His food at the home of Tapana Mishra.

When Sanatana Gosvami met Lord Chaitanya at Benares, he learned the process and principles of devotional service during two months of continuous teaching. Lord Chaitanya's instructions to Sanatana Gosvami have been described in the first part of this book. After receiving these teachings, Sanatana Gosvami was authorized to propagate the principles of devotional service and Srimad-Bhagavatam. It was during this time that both Tapana Mishra and Chandrashekhar Acharya were feeling very sorry about the strong criticism against Lord Chaitanya Mahaprabhu, and they came together and prayed for the Lord to meet the Mayavadi sannyasis.

"We have been mortified by hearing unfavorable criticisms from the Mayavadi sannyasis against You," they informed Lord Chaitanya. "Indeed, it has become intolerable for us." They requested the Lord to do something so that these criticisms might be stopped. While they were discussing this subject, a brahmana came to Lord Chaitanya and invited Him to his home. All the sannyasis had been invited but Chaitanya Mahaprabhu, and now the brahmana came to invite Him. Knowing that the Lord did not associate with Mayavadi sannyasis, the brahmana fell down at Chaitanya Mahaprabhu's feet and implored Him: "Although I know that You do not accept invitations, I still implore You to come and take prasada at my home with the other sannyasis. If You accept this invitation, I will consider it a special favour."

The Lord took this opportunity and accepted the brahmana's invitation in order to meet the Mayavadi sannyasis. Actually this was an arrangement made by the Lord Himself. Although the brahmana who invited Him knew that the Lord did not accept any invitations, he was still very eager to invite Him.

The next day Lord Chaitanya went to the house of the brahmana and saw that all the Mayavadi sannyasis were sitting there. He offered His respects to all the sannyasis as was customary, and then went to wash His feet. After washing, He sat down beside the foot basin, a little distance from the other sannyasis. While He was sitting there, the sannyasis saw a glaring effulgence emanating from His body. Being attracted by this glaring effulgence, all the Mayavadi sannyasis stood up and showed Him their respects. Amongst them was one sanyasi named Prakasananda Saraswati. He was the chief amongst the impersonalist sannyasis, and he addressed Lord Chaitanya with great humility, asking Him to come and sit amongst them.

"My dear Sir, why are You sitting in that filthy place?" he asked. "Please come and sit with us."

"Oh, I belong to an inferior sect of sannyasis," Lord Chaitanya replied. "Therefore I think that I should not sit with you. Let Me remain down here."

Prakasananda was surprised to hear such a thing from such a learned man, and he actually took the Lord's hand and requested Him to please come and sit with him. When Lord Chaitanya was finally seated amongst them, Prakasananda Saraswati said, "I think Your name is Sri Krishna Chaitanya, and I understand that You belong to our Mayavadi sect because You have taken sannyasa from Keshava Bharati, who belongs to the Shankaracharya sampradaya."

According to the Shankara sect, there are ten different names for sannyasis. Out of them, three names—Tirtha, Asrama and Saraswati—are given to the sannyasis considered to be the most enlightened and cultured. Since Lord Chaitanya was a Vaishnava, He was naturally humble and meek, and He wanted to give the better sitting place to Prakasananda, who belonged to the Saraswati sampradaya. According to Shankara's principles, a brahmacari of the Bharatic school is called Chaitanya. However, although Sri Krishna Chaitanya Mahaprabhu took sannyasa, He kept His brahmacari name and did not take up the title of Bharati.

"Well, Sir," Prakasananda Saraswati continued, "You belong to our Shankara sect, and You are living in Benares—so why don't You mix with us? What is the reason? Another thing—You are a sanyasi and are supposed to engage simply in the study of Vedanta, but we see that instead You are always engaged in chanting and dancing and playing music. What is the reason? These are the activities of emotional and sentimental people but You are a qualified sanyasi. Why not engage in the study of Vedanta? By Your effulgence it appears to us that You are just like the Supreme Narayana, the Personality of Godhead, but by Your behaviour, You appear to be otherwise. So we are inquisitive to know why You act in this way."

"My dear sir, My spiritual master considered Me a great fool," Lord Chaitanya replied. "Therefore he has more or less punished Me by saying that because I am such a fool I have no capacity to study Vedanta. So in turn he gave Me the chanting of Hare Krishna, Hare Krishna, Krishna Krishna, Hare Hare. Hare Rama, Hare Rama, Rama Rama, Hare Hare. My spiritual master told Me, 'Just go on chanting this Hare Krishna mantra: it will make You all-perfect.' "

Actually Lord Chaitanya was neither foolish nor ignorant of the principles of Vedanta. His purpose was to demonstrate to modern society that fools who have no history of penance and austerity should not try to study Vedanta just for some recreational purpose. In His Siksastaka, Lord Chaitanya said that one should be in a humble state of mind, should think himself lower than the grass on the street, should be more tolerant than a tree, and should be devoid of all sense of prestige and ready to offer all kinds of respects to others. In such a state of mind, one can chant the Vedanta philosophy or the holy name of God constantly. The Lord also wanted to teach that a serious student of transcendental science should follow the words of his spiritual master. According to the calculations of the spiritual master, Lord Chaitanya appeared to be a fool; therefore he said that He should not indulge in the study of Vedanta but should continue chanting the Hare Krishna mantra. Lord Chaitanya strictly obeyed this order. In other words, Lord Chaitanya impressed on the Mayavadis that the words of a bona fide spiritual master must be strictly followed. By following them, one becomes perfect in all respects.

Vedanta indicates that the last word of Vedic knowledge is the understanding of Krishna. Vedais ca sarvair aham eva vedyo vedanta-krd veda-vid eva caham: "By all the Vedas am I to be known; indeed, I am the compiler of Vedanta, and I am the knower of the Vedas." (Bg. 15.15) When one actually comes to understand the Vedanta, he comes to know Krishna and his relationship with Krishna. One who understands Krishna, understands everything. The knower of Krishna is always engaged in the transcendental loving service of Krishna. This is confirmed by the Lord Himself in Bhagwad Gita:

aham sarvasya prabhavo
mattah sarvam pravartate
iti matva bhajante mam
budha bhava-samanvitah

"I am the source of all spiritual and material worlds. Everything emanates from Me. The wise who know this perfectly engage in My devotional service and worship Me with all their hearts." (Bg. 10.8)

A living entity is eternally related with Krishna in the relationship of master and servant. Once that service is wanting—or, in other words, when one is not situated in Krishna consciousness—it is to be understood that study of Vedanta is insufficient. When one does not understand Krishna or does not engage in His transcendental loving service, it is to be understood that he is adverse to the study of Vedanta and to the understanding of the Supreme Personality of Godhead. The path of Vedanta study shown by Lord Chaitanya should be followed by all. A person who is puffed up by so-called education and who has no humility does not seek the protection of a bona fide spiritual master. He thinks that he does not require a spiritual master and that he can achieve the highest perfection by his own efforts. Such persons are not eligible for the study of Vedanta-sutra. Those who are under the spell of material energy do not follow the instructions of the disciplic succession but try to manufacture something of their own and thereby step outside the sphere of Vedanta study. A bona fide spiritual master must always condemn such independent mental speculators. If the bona fide spiritual master directly points out the foolishness of a disciple, it should not be taken otherwise.

A person who is completely ignorant of the science of God cannot be considered learned. More or less, everyone who is not in Krishna consciousness is subject to foolishness. Sometimes we display our foolishness by accepting someone who is barely educated as a spiritual master. It is our duty to understand the Supreme Personality of Godhead, whose lotus feet are worshipped by all the Vedas. One who does not understand Him and is proud of a false understanding of Vedanta is actually a fool. Mundane attempts at academic knowledge are simply another type of foolishness. As long as one cannot understand the cosmic manifestation as a representation of the three modes of material nature, he must be considered to be in the darkness of inebriety and caught in the duality of this material world. A person who is in perfect knowledge of Vedanta becomes a servitor of the Supreme Lord, who is the maintainer and sustainer of the whole cosmic manifestation. As long as one is not transcendental to the service of the limited, he cannot have knowledge of Vedanta.

As long as one is within the limited jurisdiction of fruitive activities or is involved in mental speculation, he may perhaps be eligible to study or teach the theoretical knowledge of Vedanta-sutra, but he cannot understand the supreme, eternal, transcendental (completely liberated) vibration of Hare Krishna, Hare Krishna, Krishna Krishna, Hare Hare. Hare Rama, Hare Rama, Rama Rama, Hare Hare. One who has achieved perfection in chanting the transcendental Hare Krishna vibration does not have to separately learn the philosophy of Vedanta-sutra.

According to the teachings of Chaitanya Mahaprabhu, the bona fide spiritual master, those who do not understand the transcendental vibration as being nondifferent from the Supreme and who try to become Mayavadi philosophers or experts in Vedanta-sutra are all fools. Studying Vedanta-sutra by one's own efforts (the ascending process of knowledge) is another sign of foolishness. He who has attained a taste for chanting the transcendental vibration, however, actually attains the conclusion of Vedanta. In this connection, there are two verses in Srimad-Bhagavatam which are very instructive. The purport of the first is that even if a low caste person is engaged in chanting the transcendental vibration, it is to be understood that he has performed all types of renunciation, austerities and sacrifice and has studied all the Brahma-sutras. Thus one can be able to chant Hare Krishna, Hare Krishna, Krishna Krishna, Hare Hare. Hare Rama, Hare Rama, Rama Rama, Hare Hare. The purport of the second verse is that one who chants the two syllables Ha-ri must be considered to have studied all the Vedas: the Rigveda, Atharva Veda, Yajur Veda and Sama Veda.

On the other hand, there are many so-called devotees who think Vedanta is not meant for devotees. Such people are ignorant of the fact that Vedanta is the only platform of pure devotees. All the great acharyas of the four Vaishnava sampradayas have made commentaries on the Vedanta-sutra, but the so-called devotees known as prakrta-sahajiya carefully avoid the study of Vedanta-sutra. The prakrta-sahajiya mistakenly take the pure devotees and Vaishnava acaryas to be mental speculators or fruitive actors. Consequently they themselves become Mayavadis and leave the service of the Supreme Lord.

Understanding the Vedanta-sutra by academic knowledge never enables one to understand the value of the transcendental vibration. People who are entangled in academic knowledge are conditioned souls who are confused about the facts of "I", "mine" and "my" understanding. Consequently they are unable to detach their minds from the external energy. When a person actually attains transcendental knowledge, he becomes free from this duality and engages in the transcendental loving service of the Supreme Lord. The Lord's service is the only means by which one can become detached from material activities. A person properly initiated by a bona fide spiritual master and engaged in chanting Hare Krishna, Hare Krishna, Krishna Krishna, Hare Hare. Hare Rama, Hare Rama, Rama Rama, Hare Hare gradually becomes freed from the conception of "I" and "mine" and becomes attached to the Lord's transcendental loving service in one of the five transcendental relationships. Such transcendental service is not a subject matter for gross and subtle bodies. Only when one can understand that there is no difference between the Supreme and His name can one be situated in Krishna consciousness. At such a time one no longer needs to make grammatical adjustments. Rather, one becomes more interested in petitioning the Lord: "Hare Krishna—O my Lord, O energy of the Lord, please engage me in Your service!"

Lord Chaitanya explained all this to Prakasananda Saraswati and told him that He had heard all this from His spiritual master. He further informed Prakasananda Saraswati that His spiritual master had taught Him that Srimad-Bhagavatam is the actual commentary on Vedanta-sutra, as stated in Srimad-Bhagavatam by Vyasadeva, the author of Vedanta-sutra.

A student is to be considered perfected when he understands the identity of the holy name and the Supreme Lord. Unless one is under the shelter of a realized spiritual master, his understanding of the Supreme is simply foolishness. However, one can fully understand the transcendental Lord by service and devotion. When Lord Chaitanya offenselessly chanted the Hare Krishna mantra, He declared that the mantra could at once deliver a conditioned soul from material contamination. In this age of Kali there is no alternative to chanting this maha-mantra. It is stated that the essence of all Vedic literature is the chanting of this holy name of Krishna: Hare Krishna, Hare Krishna, Krishna Krishna, Hare Hare. Hare Rama, Hare Rama, Rama Rama, Hare Hare. Lord Chaitanya also told Prakasananda Saraswati, "In order to convince Me about this essential fact of Vedic knowledge, My spiritual master has taught Me a verse from Brhan-naradiya Purana [38.126]. Harer nama harer nama harer namaiva kevalam. kalau nasty eva nasty eva nasty eva gatir anyatha: In this age of quarrel and hypocrisy, the only means of deliverance is the chanting of the holy name of the Lord. There is no other way. There is no other way. There is no other way."

In three out of the four millenniums (namely Satya-yuga, Treta-yuga and Dvapara-yuga) people had the honour to be able to understand transcendence through the path of disciplic succession. However, in the present age, people have no interest in the disciplic succession. Instead, they have invented many paths of logic and argument. This individual attempt to understand the supreme transcendence (called the ascending process) is not approved by the Vedas. The Absolute Truth must descend from the absolute platform. He is not to be understood by the ascending process.

The holy name of the Lord—Hare Krishna, Hare Krishna, Krishna Krishna, Hare Hare. Hare Rama, Hare Rama, Rama Rama, Hare Hare—is a transcendental vibration because it comes from the transcendental platform, the supreme abode of Krishna. Because there is no difference between Krishna and His name, the holy name of Krishna is as pure, perfect and liberated as Krishna Himself. Academic scholars have no entrance by means of logic and other argument into the understanding of the transcendental nature of the holy name of God. The single path in understanding the transcendental nature of Hare Krishna, Hare Krishna, Krishna Krishna, Hare Hare. Hare Rama, Hare Rama, Rama Rama, Hare Hare is the chanting of these names with faith and adherence. Such chanting will release one from designated conditions arising from the gross and subtle bodies.

In this age of logic, argument and disagreement, the chanting of Hare Krishna is the only means for self-realization. Because this transcendental vibration alone can deliver the conditioned soul, it is considered to be the essence of the Vedanta-sutra. According to the material conception, there is duality between the name, form, quality, emotions and activities of a person and the person himself, but as far as the transcendental vibration is concerned, there is no such limitation, for it descends from the spiritual world. In the spiritual world there is no difference between the name of the person and the quality of the person. Of course in the material world there is a difference. Because the Mayavadi philosophers cannot understand this, they cannot utter the transcendental vibration.

Lord Chaitanya then told Prakasananda Saraswati that because He received the order from His spiritual master, He was constantly chanting Hare Krishna, Hare Krishna, Krishna Krishna, Hare Hare. Hare Rama, Hare Rama, Rama Rama, Hare Hare. "As a result of this chanting," the Lord said, "I sometimes become very impatient and cannot restrain Myself from dancing and laughing or crying and singing. Indeed, I become just like a madman. When I first wondered whether I had become mad by chanting this Hare Krishna, Hare Krishna, Krishna Krishna, Hare Hare. Hare Rama, Hare Rama, Rama Rama, Hare Hare, I approached My spiritual master and informed him that I had gone mad by chanting Hare Krishna, Hare Krishna, Krishna Krishna, Hare Hare. Hare Rama, Hare Rama, Rama Rama, Hare Hare. Thus I asked him what was My actual position."

In the Narada-pancaratra it is stated:

eso vedah sad-angani
chandamsi vividhah surah
sarvam astaksarantahstham
yac canyad api vanmayam
sarva-vedanta-sararthah
samsararnava-taranah

"All Vedic rituals, mantras and understanding are compressed into eight words: Hare Krishna, Hare Krishna, Krishna Krishna, Hare Hare." Similarly, in the Kalisantarana Upanishad it is stated:

hare Krishna hare Krishna
Krishna Krishna hare hare
hare rama hare rama
rama rama hare hare

iti sodasakam namnam
kali-kalmasa-nasanam
natah parataropayah
sarva-vedesu drsyate

"The sixteen words—Hare Krishna, Hare Krishna, Krishna Krishna, Hare Hare. Hare Rama, Hare Rama, Rama Rama, Hare Hare—are especially meant for counteracting the contaminations of Kali. To save oneself from the contamination of Kali, there is no alternative but the chanting of these sixteen words."

Lord Chaitanya informed Prakasananda Saraswati that when His spiritual master understood Him, he said, "It is the transcendental nature of the holy names of Hare Krishna, Hare Krishna, Krishna Krishna, Hare Hare. Hare Rama, Hare Rama, Rama Rama, Hare Hare to transport a man into spiritual madness. Anyone who sincerely chants this holy name very soon attains elevation to the platform of love of God and becomes mad after God. This madness arising from love of God is the first perfectional stage for a human being."

Generally a human being is interested in religion, economic development, sense gratification and liberation, but love of God is above all these. A bona fide spiritual master chants the holy names—Hare Krishna, Hare Krishna, Krishna Krishna, Hare Hare. Hare Rama, Hare Rama, Rama Rama, Hare Hare—and the transcendental sound vibration enters into the ear of the disciple, and if a disciple follows in the footsteps of his spiritual master and chants the holy name with similar respect, he actually comes to worship the transcendental name. When the transcendental name is worshipped by the devotee, the name Himself spreads His glories within the heart of a devotee.

When a devotee is perfectly qualified in chanting the transcendental vibration of the holy name, he is quite fit to become a spiritual master and to deliver all the people of the world. The chanting of the holy name is so powerful that it gradually establishes its supremacy above everything in the world. The devotee who chants it becomes transcendentally situated in ecstasy and sometimes laughs, cries and dances in his ecstasy. Sometimes the unintelligent put hindrances in the path of chanting this maha-mantra, but one who is situated on the platform of love of Godhead chants the holy name loudly for all concerned. As a result, everyone becomes initiated in the chanting of the holy names—Hare Krishna, Hare Krishna, Krishna Krishna, Hare Hare. Hare Rama, Hare Rama, Rama Rama, Hare Hare. By chanting and hearing the holy names of Krishna, a person can remember the forms and qualities of Krishna.

Further Talks with Prakasananda

The transcendental ecstatic attachment for Krishna which results from perfectly understanding that Krishna's person and name are identical is called bhava. One who has attained bhava is certainly not contaminated by material nature. He actually enjoys transcendental pleasure from bhava, and when bhava is intensified, it is called love of Godhead. Lord Chaitanya told Prakasananda Saraswati that the holy name of Krishna, called the maha-mantra (great chanting), enables anyone who chants it to attain the stage of love of Godhead, or intensified bhava. Such love of Godhead is the ultimate human necessity, for when one compares it with other necessities (namely religion, economic development, sense gratification and liberation), these other so-called necessities seem most insignificant.

When one is absorbed in temporary designated existence, he hankers after sense gratification and liberation. However, love of Godhead is the eternal nature of the soul; it is unchangeable, beginningless and endless. Therefore temporary sense gratification or a desire for liberation cannot compare with the transcendental nature of love of God. Love of God is the fifth dimension in the human endeavour. Compared with the ocean of love of transcendental pleasure, the conception of impersonal Brahman is no more significant than a drop of water.

Lord Chaitanya next explained that His spiritual master had confirmed the validity of His ecstasy which resulted from His chanting the holy name of God and also confirmed that the essence of all Vedic literature is the attainment of love of Godhead. Lord Chaitanya's spiritual master had said that the

Lord was fortunate enough to have attained love of Godhead. By attaining such transcendental love, one's heart becomes very anxious to attain direct contact with the Lord. Feeling such transcendental sentiment, one sometimes laughs and sometimes cries, sings and dances like a madman, and sometimes traverses hither and thither. In this way there are various ecstatic symptoms evinced: crying, changing bodily colour, madness, bereavement, silence, pride, ecstasy and gentleness. Frequently the person in love with God dances, and such dancing places him in the ocean of the nectar of love of Krishna.

Lord Chaitanya said that His spiritual master told Him: "It is very good that You have attained such a perfectional stage of love of Godhead. Because of Your attainment, I am very much obliged to You." The father becomes more enlivened when he sees his son advance beyond himself. Similarly, the spiritual master takes more pleasure in seeing his disciple advance than in advancing himself. Thus Lord Chaitanya's spiritual master blessed Him, telling Him to "dance, sing, propagate this sankirtana movement, and by instructing people about Krishna, try to deliver them from nescience." Lord Chaitanya's spiritual master also taught Him the following verse from Srimad-Bhagavatam (11.2.40):

evam-vraṭah sva-priya-nama-kirtya
jatanurago druta-citta uccaih
hasaty atho roditi rauti gayaty
unmada-van nrtyati loka-bahyah

"A person who is constantly engaged in devotional service to Krishna and who chants His holy name becomes so transcendentally attached to the chanting that his heart becomes softened without extraneous endeavour. When this happens, he exhibits transcendental ecstasies and sometimes laughs, sometimes cries, sings and dances—not exactly in an artistic way, but just like a madman."

Lord Chaitanya further informed Prakasananda Saraswati: "Because I have full faith in My spiritual master's words, I am always engaged in chanting Hare Krishna, Hare Krishna, Krishna Krishna, Hare Hare. Hare Rama, Hare Rama, Rama

Rama, Hare Hare. I do not exactly know how I have become just like a madman, but I believe the name of Krishna has induced Me. I realize that the transcendental pleasure derived from chanting Hare Krishna, Hare Krishna, Krishna Krishna, Hare Hare. Hare Rama, Hare Rama, Rama Rama, Hare Hare is just like an ocean. In comparison, all other pleasures, including the pleasure of impersonal realization, are like shallow water in channels."

It appears from the talks of Lord Chaitanya that a person who cannot keep his faith in the words of the spiritual master and who acts independently cannot attain the desired success in chanting Hare Krishna. In the Vedic literatures it is stated that the import of all transcendental literature is revealed to one who has unflinching faith in the Supreme Lord and his spiritual master. Lord Chaitanya firmly believed in the statements of His spiritual master, and He never neglected the instructions of His spiritual master by stopping His sankirtana movement. Thus the transcendental potency of the holy name encouraged Him more and more in chanting Hare Krishna, the maha-mantra.

Lord Chaitanya immediately informed Prakasananda that in the modern age people in general are more or less bereft of all spiritual intellect. When such people come under the influence of Shankaracharya's Mayavadi (impersonalist) philosophy before beginning the most confidential Vedanta-sutras, their natural tendency toward obedience to the Supreme is checked. The supreme source of everything is naturally respected by everyone, but this natural tendency is hampered when one takes to the impersonalist conceptions of Shankara.

Thus the spiritual master of Lord Chaitanya suggested that it is better that one not study the Sariraka-bhasya of Shankaracharya, for it is very harmful to people in general. Indeed, the common man does not even have the intelligence to penetrate into the jugglery of words. He is better advised to chant the maha-mantra: Hare Krishna, Hare Krishna, Krishna Krishna, Hare Hare. Hare Rama, Hare Rama, Rama Rama, Hare Hare. In this quarrelsome age of Kali there is no alternative for self-realization.

After hearing the arguments and talks of Chaitanya Mahaprabhu, all the Mayavadi sannyasis who were present became pacified and replied with sweet words: "Dear sir, what You have spoken is all true. A person who attains love of Godhead is certainly very fortunate, and undoubtedly You are very fortunate to have attained this stage. But what is the fault in Vedanta? It is the duty of a sanyasi to read and understand Vedanta. Why do You not study it?"

According to Mayavadi philosophers, Vedanta refers to the Sariraka commentary of Shankaracharya. When impersonal philosophers refer to Vedanta and the Upanisads, they are actually referring to the commentaries of Shankaracharya, the greatest teacher of Mayavadi philosophy. After Shankaracharya came Sadananda-yogi, who claimed that the Vedanta and Upanisads should be understood through the commentaries of Shankaracharya. Factually, this is not so. There are many commentaries on Vedanta and the Upanisads made by the Vaishnava acharyas, and these are preferred to those of Shankaracharya. However, the Mayavadi philosophers influenced by Shankaracharya do not attribute any importance to the Vaishnava understandings. There are four different sects of Vaishnava acharyas—the Suddhadvaita, Vishistadvaita, Dvaitadvaita and Acintya-bhedabheda. All the Vaishnava acharyas in these schools have written commentaries on the Vedanta-sutra, but the Mayavadi philosophers do not recognize them. The Mayavadis distinguish between Krishna and Krishna's body, and therefore they do not recognize the worship of Krishna by the Vaishnava philosophers. Thus when the Mayavadi sannyasis asked Lord Chaitanya why He did not study the Vedanta-sutra, the Lord replied, "Dear sirs, you have asked why I do not study Vedanta, and in answer to this I would speak something, but I am afraid that you would be sorry to hear it."

"We shall be very much pleased to hear You," all the sannyasis replied. "You appear just like Narayana, and Your speeches are so nice that we are taking great pleasure in them. We are very much obliged to see and hear You. Therefore we shall be very glad to hear patiently and accept whatever You say."

The Lord then began to speak on Vedanta philosophy as follows: Vedanta-sutra is spoken by the Supreme Lord Himself. The Supreme Lord, in His incarnation as Vyasadeva, has compiled this great philosophical treatise. Since Vyasadeva is an incarnation of the Supreme Lord, he cannot be likened to an ordinary person, who has the four defects which arise due to contact with material existence. The defects of a conditioned soul are: (1) he must commit mistakes; (2) he must be illusioned; (3) he must possess the tendency to cheat others; and (4) all his senses must be imperfect. We must understand that the incarnation of God is transcendental to all these defects.

Thus whatever has been spoken and written by Vyasadeva is considered to be perfect. The Upanisads and Vedanta-sutra aim at the same goal: the Supreme Absolute Truth. When we accept the import of Vedanta-sutra and the Upanisads directly as they are stated, we become glorified. The commentaries made by Shankaracharya, however, are indirect and are very dangerous for the common man to read, for by understanding the import of the Upanisads in such an indirect, disruptive way, one practically bars himself from spiritual realization.

According to the Skanda and Vayu Puranas, the word sutra refers to a condensed work which carries meaning and import of immeasurable strength without mistake or fault. The word vedanta means "the end of Vedic knowledge." In other words, any book which deals with the subject matter indicated by all the Vedas is called Vedanta. For example, Bhagwad Gita is Vedanta because in Bhagwad Gita the Lord says that the ultimate goal of all Vedic research is Krishna. Thus Bhagwad Gita and Srimad-Bhagavatam, which aim only at Krishna, are to be understood to be Vedanta.

In transcendental realization there are three divisions of knowledge called prasthana-traya. That department of knowledge which is proved by Vedic instruction (like the Upanisads) is called sruti-prasthana. Authoritative books indicating the ultimate goal and written by liberated souls like Vyasadeva (for example, Bhagwad Gita, Mahabharata and the Puranas, especially Srimad-Bhagavatam, the Maha-purana) are called smriti-prasthana. From Vedic literatures we understand that the Vedas originated from the breathing of Narayana. Vyasadeva, who is an incarnation of the power of Narayana, has compiled the Vedanta-sutra (nyaya-prasthana), but according to Shankara's commentaries, Apantaratama Rsi is also accredited with having compiled the codes of Vedanta-sutra.

According to Lord Chaitanya, the codes of the Pancaratra and the codes of Vedanta are one and the same. Since the Vedanta-sutra is compiled by Vyasadeva, it should be understood to be spoken by Narayana Himself. From all descriptive literatures dealing with Vedanta-sutra, it appears that there were many other rsis contemporary with Vyasadeva who also discussed Vedanta-sutra. These sages were Atreya, Asmarathya, Audulomi, Karsnajini, Kasakrtsna, Jaimini, Badari and other sages such as Parasari and Karmandi.

Actually in the first two chapters of Vedanta-sutra the relationship between the living entities and the Supreme Lord is explained, and in the Third Chapter the discharge of devotional service is explained. The Fourth Chapter deals with the relationship which results from discharging devotional service. The natural commentary on Vedanta-sutra is Srimad-Bhagavatam. The great acharyas of the four Vaishnava communities (sampradayas)—namely, Ramanujacarya, Madhvacarya, Vishnusvami and Nimbarka—have also written commentaries on Vedanta-sutra by following the principles of Srimad-Bhagavatam.

At present the followers of all the acharyas have written many books following the principles of Srimad-Bhagavatam as the commentary on the Vedanta. Shankara's commentary on Vedanta-sutra, known as Sariraka-bhasya, is very much adored by the impersonalist scholars, but commentaries written on the Vedanta written from the materialistic point of view are completely adverse to the transcendental service of the Lord. Consequently Lord Chaitanya said that direct commentaries on the Upanisads and Vedanta-sutra are glorious, but that anyone who follows the indirect path of Shankaracharya's Sariraka-bhasya is certainly doomed.

Lord Chaitanya admitted that Shankaracharya was an incarnation of Lord Shiva, and it is known that Lord Shiva is one of the greatest devotees (a mahajana) of the Bhagavata school. There are twelve great authorities on devotional service, and Lord Shiva is one of them. Why, then, did he adopt the process of Mayavadi philosophy? The answer is given in Padma Purana, where Lord Shiva states:

mayavadam asac-chastram
pracchannam bauddham ucyate
mayaiva kalpitam devi
kalau brahmana-rupina

"The Mayavadi philosophy is veiled Buddhism." In other words, the voidist philosophy of Buddha is more or less repeated in the Mayavadi philosophy of impersonalism, although the Mayavadi philosophy claims to be directed by the Vedic conclusions. Lord Shiva, however, admits that this philosophy is manufactured by him in the age of Kali in order to mislead the atheists. "Actually the Supreme Personality of Godhead has His transcendental body," Lord Shiva states. "But I describe the Supreme as impersonal. I also explain the Vedanta-sutra according to the same principles of Mayavadi philosophy."

In the Shiva Purana the Supreme Lord says:

dvaparadau yuge bhutva
kalaya manusadisu
svagamaih kalpitais tvam ca
janan mad-vimukhan kuru

"In the beginning of the Dvapara-yuga, directed by My orders, many sages will bewilder the people in general by Mayavadi philosophy." In the Padma Purana Lord Shiva personally tells Bhagavatidevi:

srnu devi paraksyami
tamasani yatha-kramam
yesam sravana-matrena
patityam jnaninam api
apartham sruti-vakyanam
darsayal loka-garhitam

karma-svarupa-tyajyatvam
atra ca pratipadyate

sarva-karma-paribhramsan
naiskarmyam tatra cocyate
paratma jivayor aikyam
mayatra pratipadyate

"My dear Devi, sometimes I teach Mayavadi philosophy for those who are engrossed in the mode of ignorance. But if a person in the mode of goodness happens to hear this Mayavadi philosophy, he falls down, for when teaching Mayavadi philosophy, I say that the living entity and the Supreme Lord are one and the same."

Sadananda-yogi, one of the greatest Mayavadi acharyas, has written in his book, Vedanta-sara: "The Absolute Truth of eternity, knowledge and bliss is Brahman. Ignorance and all products of ignorance are non-Brahman. All products of the three modes of material nature are covered by ignorance, and all are different from the supreme cause and effect. This ignorance is manifested in a collective and individual sense. Collective ignorance is called visuddha-sattva-pradhana. When that visuddha-sattva-pradhana is manifest within the ignorance of material nature, it is called the Lord, and the Lord manifests all kinds of ignorance. Therefore He is known as sarvajna." Thus according to Mayavadi philosophy, the Lord is a product of this material nature, and the living entity is in the lowest stage of ignorance. That is the sum and substance of Mayavadi philosophy.

If, however, we accept the import of the Upanisads directly, it is clear that the Supreme Personality of Godhead is a person with unlimited potency. For example, in the Svetasvatara Upanishad it is stated that the Supreme Personality of Godhead is the origin of everything and that He has multiple potencies. The Supreme Personality of Godhead is transcendental to the cosmic manifestation. He is the origin of all religion, the supreme deliverer and the possessor of all opulences. The Supreme Personality of Godhead, who is just like the sun, profusely distributes His energies while situated beyond the cloud of this material cosmic manifestation. He is the master of masters, and He is the Supreme of the supremes. He is known as the greatest Lord, the Personality of Godhead. His energies and potencies are multiple and variously distributed. It is also stated that Vishnu is the Supreme and that saintly persons are always anxious to see His lotus feet (Rg Veda 1.22.20). In the Aitareya Upanishad it is also stated that the cosmic manifestation came about when the Lord glanced over material nature (1.1.1-2). This is also verified by the Prasna Upanishad (6.3).

In the negative descriptions of the Lord which occur in Vedic literature (as in apani-padah) there are indications that the Lord has no material body and no material form. However, He does have His spiritual transcendental body and His transcendental form. Because the Mayavadi philosophers misunderstand His transcendental nature, they explain Him as impersonal. The Lord's name, form, quality, entourage and abode are all in the transcendental world. How can He be a transformation of this material nature? Everything connected with the Supreme Lord is eternal, blissful and full of knowledge.

In effect, Shankaracharya preached Mayavadi philosophy in order to bewilder a certain type of atheist. Actually he never considered the Supreme Lord, the Personality of Godhead, to be impersonal or to have no body or form. It is best for intelligent persons to avoid lectures on Mayavadi philosophy. We should understand that the Supreme Personality of Godhead Vishnu is not impersonal. He is a transcendental person, and the basic principle of the cosmic manifestation is His energy. Mayavadi philosophy cannot trace the energy of the Supreme Lord, but all Vedic literatures give evidence of the Supreme Lord's various energetic manifestations.

Vishnu is not a product of material nature, but material nature is a product of Vishnu's potency. The Mayavadi philosophers understand Vishnu to be a product of material nature, but if Vishnu is a product of material nature, He can only be counted amongst the demigods. One who considers Vishnu to be a demigod is certainly mistaken and misled. How this is so is explained in Bhagwad Gita: "Deluded by the three modes, the whole world does not know Me who am above the modes and inexhaustible. This divine energy of Mine, consisting of the three modes of material nature, is difficult to overcome, but those who have surrendered unto Me can easily cross beyond it." (Bg. 7.13-14)

The Goal of Vedanta Study

It is concluded that Lord Krishna, or Vishnu, is not of this material world. He belongs to the spiritual world. One who considers Him to be a material demigod is a great offender and blasphemer. Lord Vishnu is not subject to perception by material senses, nor can He be realized by mental speculation. There is no difference between the body and soul of the Supreme Lord Vishnu, although in the material world there is always a difference between the body and the soul.

Things material are enjoyed by the living entities because the living entities are superior, whereas material nature is of inferior quality. Thus the superior quality, the living entities, can enjoy the inferior quality, matter. Because Lord Vishnu is in no way touched by matter, He is not subject to enjoy material nature the way the living entities do. The living entities cannot attain knowledge of Vishnu by enjoying their habits of mental speculation. The infinitesimal living entities are not the enjoyers of Vishnu, but they are enjoyed by Vishnu. Only the greatest offender thinks that Vishnu is enjoyed. The greatest blasphemy is to consider Vishnu and the living entity on the same level.

The Supreme Absolute Truth, the Personality of Godhead, is compared to a blazing fire, and the innumerable living entities are compared to sparks emanating from that fire. Although both the Supreme Lord and the living entities are qualitatively fire, there is yet a distinction. Vishnu the Supreme is infinite, whereas the living entities, which are but sparks, are infinitesimal. The infinitesimal living entities are emanations from the original infinite spirit. In their constitutional position as infinitesimal spirits, there is no trace of matter.

The living entities are not as great as Narayana, Vishnu, who is beyond this material creation. Even Shankaracharya accepts Narayana to be beyond the material creation. Since neither Vishnu nor the living entity are of the material creation, someone may inquire, "Why were the small particles of spirit created at all?" The answer is that the Supreme Absolute Truth is complete in His perfection when He is both infinite and infinitesimal. If He is simply infinite and is not infinitesimal, He is not perfect. The infinite portion is the Vishnu-tattva, or the Supreme Personality of Godhead, and the infinitesimal portion is the living entity.

Due to the infinite desires of the Supreme Personality of Godhead, there is existence in the spiritual world, and due to the infinitesimal desires of the living entity, there is existence in the material world. When the infinitesimal living entities are engaged in their infinitesimal desires for material enjoyment, they are called jiva-shakti, but when they are dovetailed with the infinite, they are called liberated souls. There is no need to ask, therefore, why God created the infinitesimal portions; they are simply the complementary side of the Supreme. It is doubtlessly essential for the infinite to have infinitesimal portions which are inseparable parts and parcels of the supreme soul. Because the living entities are infinitesimal parts and parcels of the Supreme, there is a reciprocation of feelings between the infinite and the infinitesimal.

Had there been no infinitesimal living entities, the Supreme Lord would have been inactive, and there would not be variegatedness in spiritual life. There is no meaning to a king if there are no subjects, and there is no meaning to the Supreme God if there are no infinitesimal living entities. How can there be meaning to the word "lord" if there is no one to overlord? The conclusion is that the living entities are considered to be expansions of the energy of the Supreme Lord, and the Supreme Lord, the Personality of Godhead, Krishna, is the energetic.

In all Vedic literatures, including Bhagwad Gita and Vishnu purana, much evidence is given to distinguish between the energy and the energetic. In Bhagwad Gita (Bg. 7.4) it is clearly stated that earth, water, fire, air and ether are the five principal gross elements of the material world and that mind, intelligence and false ego are the three subtle elements. All material nature is divided into these eight elements which together comprise the inferior nature, or energy, of the Lord. Another name for this inferior nature is maya, or illusion. Beyond these eight inferior elements there is a superior energy, which is called para-prakrti. That para-prakrti is the living entity, who is found in great numbers throughout the material world. He is indicated in Bhagwad Gita (Bg. 7.5) as jiva-bhutam. The purport is that the Supreme Lord is the Absolute Truth, the energetic, and as such He has His energies. When His energy is not properly manifested, or when it is covered by some shadow, it is called maya-shakti. The material cosmic manifestation is a product of that covered maya-shakti.

The living entities are factually beyond this covered inferior energy. They have their pure spiritual existence and their pure identity, as well as their pure mental activities. All of them are beyond the manifestation of this material cosmos. Although the living entity's mind, intelligence and identity are beyond the range of this material world, when he enters into this material world due to his desire to dominate matter, his original mind, intelligence and body become covered by the material energy. When he is again uncovered from these material or inferior energies, he is called liberated.

When he is liberated, he has no false ego, but his real ego again comes into existence. Foolish mental speculators think that after liberation one's identity is lost, but that is not so. Because the living entity is eternally part and parcel of God, when he is liberated, he revives his original, eternal, part-and-parcel identity.

The realization of aham brahmasmi ("I am not this body") does not mean that the living entity loses his identity. At the present moment a person may consider himself to be matter, but in his liberated state he will understand that he is not matter but spirit soul, part of the infinite. To become Krishna conscious or spiritually conscious and to engage in the transcendental loving service of Krishna are signs of the liberated stage. In the Vishnu Purana (6.7.61) it is clearly stated:

Vishnu-shaktih para prokta
ksetra-jnakhya tatha para

avidya-karma-samjnanya
trtiya shaktir isyate

"The energy of the Supreme Lord is divided into three: para, ksetrajna and avidya." The para energy is actually the energy of the Supreme Lord Himself; the ksetrajna energy is the living entity; and the avidya energy is the material world, or maya. It is called avidya, or ignorance, because under the spell of this material energy one forgets his actual position and his relationship with the Supreme Lord. The conclusion is that the living entities represent one of the energies of the Supreme Lord, and as infinitesimal parts and parcels of the Supreme, they are called jivas. If the jivas are artificially placed on the same level with the infinite Supreme—for both of them are Brahman, or spirit—bewilderment will certainly be the result.

Generally Mayavadi philosophers are perplexed before a learned Vaishnava because the Mayavadis cannot explain the cause of bondage of the living entities. They simply say, "It is due to ignorance," but they cannot explain why the living entities are covered by ignorance if they are supreme. The actual reason is that the living entities, although qualitatively one with the Supreme, are infinitesimal, and not infinite. Had they been infinite, there would have been no possibility of their being covered by ignorance. Because the living entity is infinitesimal, he is covered by an inferior energy. The foolishness and ignorance of the Mayavadis are revealed when they try to explain how it is the infinite is covered by ignorance. It is offensive to attempt to qualify the infinite as being subject to the spell of ignorance.

Although Shankara was attempting to cover the Supreme Lord by his Mayavadi philosophy, he was simply following the order of the Supreme Lord. It should be understood that his teachings were a timely necessity but not a permanent fact. In the Vedanta-sutra the distinction between the energy and the energetic is accepted from the very beginning. In that Vedanta-sutra the first aphorism (janmady asya) clearly explains that the Supreme Absolute Truth is the origin or source of all emanations. Thus the emanations are the energy of the Supreme, whereas the Supreme Himself is the energetic.

Shankara has falsely argued that if the transformation of energy is accepted, the Supreme Absolute Truth cannot remain immutable. But this is not true. Despite the fact that unlimited energy is always being generated, the Supreme Absolute Truth remains always the same. He is not affected by the emanation of unlimited energies. Shankaracharya has therefore incorrectly established his theory of illusion.

Ramanujacarya has discussed this point very nicely: "If you argue that before the creation of this material world there was only one Absolute Truth, then how is it possible that the living entity emanated from Him? If He were alone, how could He have produced or generated the infinitesimal living entities?" In answer to this question, the Vedas state that everything is generated from the Absolute Truth, everything is maintained by the Absolute Truth, and, after annihilation, everything enters into the Absolute Truth. From this statement it is clear that the living entities enter into the supreme existence when they are liberated, and they do not change their original constitutional position.

We must always remember that the Supreme Lord has His creative function and that the infinitesimal living entities have their creative functions also. It is not that their creative function is lost when they are liberated and enter into the Supreme after the dissolution of the material body. On the contrary, the creative function of the living entity is properly manifested in the liberated state. If the living entity's activities are manifest even when he is materially conditioned, then how is it possible for his activities to stop when he attains liberation? The living entity's entering the state of liberation may be compared to a bird entering a tree, or an animal entering the forest, or a plane entering the sky. In no case is identity lost.

When explaining the first aphorism of the Vedanta-sutra, Shankara most unceremoniously tried to explain that Brahman, or the Supreme Absolute Truth, is impersonal. He also cunningly tried to switch the doctrine of by-product into the doctrine of change. For the Supreme Absolute Truth, there is no change. It is simply that a by-product results from His inconceivable powers of action. In other words, a relative truth is produced out of the Supreme Truth. When a chair is produced out of crude wood, it is said that a by-product is produced.

The Supreme Absolute Truth, Brahman, is immutable, and when we find a by-product—the living entity or this cosmic manifestation—it is a transformation, or a by-product of the Supreme. It is like milk being transformed into yogurt. In this way, if we study the living entities in the cosmic manifestation, it will appear that they are not different from the original Absolute Truth, but from Vedic literatures we understand that the Absolute Truth has varieties of energy and that the living entities and the cosmic manifestation are but a demonstration of His energies. The energies are not separate from the energetic; therefore the living entity and cosmic manifestation are inseparable truths, part of the Absolute Truth. Such a conclusion regarding the Absolute Truth and the relative truth should be acceptable to any sane man.

The Supreme Absolute Truth has His inconceivable potency, out of which this cosmos has been manifested. In other words, the Supreme Absolute Truth is the ingredient, and the living entity and cosmic manifestation are the by-products. In the Taittiriya Upanishad it is clearly stated, yato va imani bhutani jayante: "The Absolute Truth is the original reservoir of all ingredients, and this material world and its living entities are produced from those ingredients."

Unintelligent persons who cannot understand this doctrine of byproducts cannot grasp how the cosmic manifestation and the living entity are simultaneously one and different from the Absolute Truth. Not understanding this, one concludes, out of fear, that this cosmic manifestation and the living entity are false. Shankaracharya gives the example of a rope being mistaken for a snake, and sometimes the example of mistaking an oyster shell for gold is cited, but surely such arguments are ways of cheating. As mentioned in the Mandukya Upanishad, the rope for a snake and the oyster for gold examples have their different applications and can be understood as follows. The living entity in his original constitutional position is pure spirit. When a human being identifies himself with the material body, he may be said to be mistaking a rope for a snake, or an oyster shell for gold. The doctrine of transformation is accepted when one thing is mistaken for another. Actually the body is not the living entity, but the doctrine of transformation accepts the body as the living entity. Every conditioned soul is undoubtedly contaminated by this doctrine of transformation.

The conditional state of the living entity is his diseased condition. Originally the living entity and the original cause of this cosmic manifestation exist outside the state of transformation. However, mistaken thoughts and arguments can overcome a person when he forgets the inconceivable energies of the Supreme Lord. Even in the material world there are many examples. The sun has been producing unlimited energy from time immemorial, and so many by-products result from the sun; yet there is no change in the heat and temperature of the sun itself. Despite its being a material product, if the sun can maintain its original temperature and yet produce so many byproducts, is it difficult for the Supreme Absolute Truth to remain unchanged in spite of producing so many by-products by His inconceivable energy? Thus there is no question of transformation as far as the Supreme Absolute Truth is concerned.

In Vedic literatures there is information of a material product called "touchstone" which simply by touch can transform iron into gold. The touchstone can produce an unlimited quantity of gold and yet remain the same. Only in the state of ignorance can one accept the Mayavadi conclusion that this cosmic manifestation and the living entities are false or illusory. No sane man would impose ignorance and illusion upon the Supreme Absolute Truth, who is absolute in everything. There is no possibility of change, ignorance or illusion being in Him. The Supreme Brahman is transcendental and completely different from all material conceptions. In the Supreme Absolute Truth there is every possible inconceivable energy existing. In the Svetasvatara Upanishad it is stated that the Supreme Absolute Personality of Godhead is full of inconceivable energies and that no one else possesses such energies.

By misunderstanding the inconceivable energies of the Supreme, one may falsely conclude that the Supreme Absolute

Truth is impersonal. Such a deluded conclusion is experienced by a living being when he is in an acute stage of disease. In Srimad-Bhagavatam also there are statements to the effect that the supreme atma, the Lord, has inconceivable and innumerable potencies. (Bhag. 3.33.3) It is also stated in Brahma-sutra that the supreme spirit has many variegated and inconceivable energies. Nor should one think that there is any possibility of ignorance existing in the Absolute Truth. Ignorance and knowledge are conceptions in this world of duality, but in the Absolute there is no duality.

It is simply foolishness to consider that the Absolute is covered by ignorance. If the Absolute Truth can possibly be covered by ignorance, how can it be said to be Absolute? Understanding the inconceivability of the Absolute is the only solution to the question of duality. This is because duality arises from the inconceivable energy of the Absolute. By His inconceivable energies, the Supreme Absolute Truth can remain unchanged and yet produce this cosmic manifestation with all its living entities, just as touchstone can produce unlimited quantities of gold and yet remain unchanged. Because the Absolute Truth has such inconceivable energies, the material quality of ignorance cannot pertain to Him.

The true variegatedness which exists in the Absolute Truth is a product of His inconceivable energy. Indeed, it can be safely concluded that this cosmic manifestation is but a by-product of His inconceivable energies. Once we accept the inconceivable energies of the Supreme Lord, we will find that there is no duality at all. The expansion of the energy of the Supreme Lord is as true as the Supreme Lord. As far as the manifestation of the supreme energy is concerned, there is no question of transformation. The same example can be cited: in spite of producing unlimited quantities of gold, the touchstone remains the same. We therefore hear some sages say that the Supreme is the ingredient or cause of this cosmic manifestation.

Actually the example of the rope and the snake is not completely irregular. When we accept a rope to be a snake, it is to be understood that we have experienced a snake previously. Otherwise, how can the rope be mistaken for a snake? Thus the conception of a snake is not untrue or unreal in itself. It is the false identity that is untrue or unreal. When, by mistake, we consider the rope to be a snake, that is our ignorance. But the very idea of a snake is not in itself ignorance. When we accept a mirage to be water in the desert, there is no question of water being a false concept. Water is a fact, but it is a mistake to think that there is water in the desert.

Thus this cosmic manifestation is not false, as Shankaracharya maintains. Actually there is nothing false here. The Mayavadis say that this world is false because of their ignorance. It is the conclusion of Vaishnava philosophy that this cosmic manifestation is a by-product of the inconceivable energies of the Supreme Lord.

The principal word in the Vedas, pranava omkara, is the sound representation of the Supreme Lord. Therefore omkara should be considered the supreme sound. However, Shankaracharya has falsely preached that tat tvam asi are the supreme vibrations. Omkara is the reservoir of all the energies of the Supreme Lord. Shankara is wrong in maintaining that the words tat tvam asi are the supreme vibrations of the Vedas, for tat tvam asi are secondary words only. tat tvam asi suggests only a partial representation.

In Bhagwad Gita the Lord has in many places given importance to omkara, (Bg. 8.13, 9.17, 17.24). Similarly, omkara is given importance in the Atharva Veda and the Mandukya Upanishad. In his Bhagavat-sandarbha, Srila Jiva Gosvami says: "Omkara is the most confidential sound representation of the Supreme Lord." The sound representation or name of the Supreme Lord is as good as the Supreme Lord Himself. By vibrating the sound of omkara, or of Hare Krishna, Hare Krishna, Krishna Krishna, Hare Hare. Hare Rama, Hare Rama, Rama Rama, Hare Hare, one can be delivered from the contamination of this material world. Because such vibrations of transcendental sound can deliver a conditioned soul, they are known as tara, or deliverers.

That the sound vibration of the Supreme Lord is identical with the Supreme Lord is a fact. This is confirmed in the Narada-pancaratra:

vyaktam hi Bhagvan eva
saksan-narayanah svayam
astaksara-svarupena
mukhesu parivartate

"When the transcendental sound vibration is practiced by a conditioned soul, the Supreme Lord is present on his tongue." In the Mandukya Upanishad it is said that when omkara is chanted, whatever is seen as material is seen perfectly as spiritual. In the spiritual world or in spiritual vision there is nothing but omkara, or the one alternate, om. Unfortunately, Shankara has abandoned this chief word, omkara, and has whimsically accepted tat tvam asi as the supreme vibration of the Vedas. By accepting such a secondary word and leaving aside the principal vibration, he has given up a direct interpretation of the scripture in favour of his own indirect interpretation.

Sripada Shankaracharya has unceremoniously obscured the Krishna consciousness described in the purusa-vedanta-sutra by manufacturing an indirect interpretation and abandoning the direct interpretation. Unless we take all the statements of Vedanta-sutra as self-evident, there is no point in studying Vedanta-sutra. Interpreting the verses of Vedanta-sutra according to one's own whim is the greatest disservice to the self-evident Vedas.

As far as the omkara pranava is concerned, it is considered to be the sound incarnation of the Supreme Personality of Godhead. As such, omkara is eternal, unlimited, transcendental, supreme and indestructible. He (omkara) is the beginning, middle and end, and He is beginningless as well. When one understands omkara as such, he becomes immortal. One should thus know omkara as a representation of the Supreme situated in everyone's heart. One who understands omkara and Vishnu as being one and the same and all-pervading never laments in the material world, nor does he remain a sudra.

Although He (omkara) has no material form, He is unlimitedly expanded, and He has unlimited form. By understanding omkara one can become free from the duality of the material world and attain absolute knowledge. Therefore omkara is the most auspicious representation of the Supreme Lord. Such is the description given by Mandukya Upanishad.

One should not foolishly interpret an Upanisadic description and say that because the Supreme Personality of Godhead "cannot" appear Himself in this material world in His own form, He sends His sound representation (omkara) instead. Due to such a false interpretation, omkara comes to be considered something material, and consequently omkara is misunderstood and praised as being simply an exhibition or symbol of the Lord. Actually omkara is as good as any other incarnation of the Supreme Lord.

The Lord has innumerable incarnations, and omkara is one of them. As Krishna states in Bhagwad Gita: "Amongst vibrations, I am the syllable om." (Bg. 9.17) This means that omkara is nondifferent from Krishna. Impersonalists, however, give more importance to omkara than to the Personality of Godhead, Krishna.

The fact is, however, that any representational incarnation of the Supreme Lord is nondifferent from Him. Such an incarnation or representation is as good spiritually as the Supreme Lord. Omkara is therefore the ultimate representation of all the Vedas. Indeed, the Vedic mantras or hymns have transcendental value because they are prefixed by the syllable om. The Vaishnavas interpret omkara as follows: by the letter O, Krishna, the Supreme Personality of Godhead, is indicated; by the letter U, Krishna's eternal consort Srimati Radharani is indicated; and by the letter M, the eternal servitor of the Supreme Lord, the living entity, is indicated.

Shankara has not given such importance to the omkara. However, importance is given in the Vedas, the Ramayana, the Puranas and in the Mahabharata from beginning to end. Thus the glories of the Supreme Lord, the Supreme Personality of Godhead, are declared.

3

The Srimad Bhagavatam

After the conversion of the Mayavadi sannyasis to the path of Chaitanya Mahaprabhu, many scholars and inquisitive people visited the Lord at Benares. Since it was not possible for everyone to see Chaitanya Mahaprabhu at His residence, people used to stand in line to see Him as He passed on His way to the temple of Vishvanatha and Bindumadhava. One day, when the Lord visited the temple with His associates, Chandrashekhar Acharya, Paramananda, Tapana Mishra, Sanatana Gosvami and others, He was singing:

haraye namah Krishna yadavaya namah
gopala govinda rama sri-madhusudana

When the Lord sang in this way, chanting and dancing, thousands of people gathered around Him, and when the Lord chanted, they roared. The vibration was so tumultuous that Prakasananda Saraswati, who was sitting nearby, immediately joined the crowd with his disciples. As soon as he saw the beautiful body of Lord Chaitanya and the Lord dancing with His associates, Prakasananda Saraswati also joined and began to sing: "Hari! Hari!" All the inhabitants of Benares were struck with wonder upon seeing the ecstatic dancing of Lord Chaitanya.

However, Lord Chaitanya checked His continuous ecstasy and stopped dancing when He saw the Mayavadi sannyasis. As soon as the Lord stopped chanting and dancing, Prakasananda Saraswati fell at His feet. Trying to stop him, Lord Chaitanya said, "Oh, you are the spiritual master of the whole world, jagad-guru, and I am not even equal to your disciples. You should therefore not worship an inferior like Me.

You are exactly like the Supreme Brahman, and if I allow you to fall down at My feet, I will commit a very great offense. Although you have no vision of duality, for the sake of teaching the people in general you should not do this."

"Previously I spoke ill of You many times," Prakasananda Saraswati replied. "Now in order to free myself from the results of my offense, I fall down at Your feet." He then quoted a verse from Vedic literatures which states that when even a liberated soul commits an offense against the Supreme Lord, he again becomes a victim of material contamination. Prakasananda Saraswati then quoted another verse from Srimad-Bhagavatam (10.34.9) regarding Nanda Maharaja's being attacked by a serpent who was previously VidyadhNrarcita. When the serpent was touched by the lotus feet of Krishna, he regained his previous body and was freed from the reactions of his sinful activities.

When Lord Chaitanya thus heard Himself equated with Krishna, He mildly protested. He wanted to warn people in general not to compare the Supreme Lord with any living entity, and although He was the Supreme Lord Himself, He protested against this comparison in order to teach us. Thus He said that it is the greatest offense to compare anyone with the Supreme Lord Krishna. Lord Chaitanya always maintained that Vishnu, the Supreme Personality of Godhead, is great and the living entities, however great they may be, are but infinitesimal. In this connection, He quoted a verse from Padma Purana which is found in the Vaishnava tantra (Hari-bhakti-vilasa 1.73): "A person who compares the Supreme Lord even with the greatest of demigods like Brahma and Shiva must be considered a number one atheist."

"I can understand that You are the Supreme Personality of Godhead, Krishna," Prakasananda Saraswati continued, "and even though You present Yourself as a devotee, You are still worshipable because You are greater than all of us in education and realization. Therefore by blaspheming You, we have committed the greatest offense. Please excuse us."

How a devotee becomes the greatest of all transcendentalists is stated in Srimad-Bhagavatam (6.14.5):

muktanam api siddhanam
narayana-parayanah
sudurlabhah prasantatma
kotisv api mahamune

"There are many liberated souls and perfected souls, but out of all of them he who is a devotee of the Supreme Personality of Godhead is best. The devotees of the Supreme Lord are always calm and quiet, and their perfection is very rarely seen, even in millions of persons." Prakasananda also quoted another verse (Bhag. 10.4.46) in which it is stated that one's duration of life, prosperity, fame, religion and the benediction of higher authorities are all lost when one offends a devotee. Prakasananda also quoted another verse from Srimad-Bhagavatam (7.5.32) which said that all the misgivings of the conditioned soul disappear at the touch of the lotus feet of the Supreme Personality of Godhead.

However, one cannot touch the lotus feet of the Supreme Lord unless one receives the benediction of the dust of the lotus feet of the Lord's pure devotee. In other words, one cannot become a pure devotee of the Supreme Personality of Godhead unless he is favoured by another pure devotee of the Lord.

"Now I am taking shelter of Your lotus feet," Prakasananda Saraswati said, "for I want to be elevated to the position of a devotee of the Supreme Lord."

After talking in this way, both Prakasananda Saraswati and Lord Chaitanya sat together. "Whatever You have said concerning discrepancies in the Mayavadi philosophy is also known by us," Prakasananda said. "Indeed, we know that all the commentaries on Vedic scriptures by Mayavadi philosophers are erroneous, especially those of Shankaracharya. Shankaracharya's interpretations of Vedanta-sutra are all figments of his imagination. You have not explained the codes of the Vedanta-sutra and Upanisads according to Your own imagination but have presented them as they are. Thus we are all pleased to have heard Your explanation. Such explanations of the codes of Vedanta-sutra and the Upanisads cannot be given by anyone but the Supreme Personality of Godhead. Since You have all the potencies of the Supreme Lord, please explain the Vedanta-sutra further so that I may be benefited." Lord Chaitanya protested against being called the Supreme Lord, and He said, "My dear sir, I am an ordinary living entity. I cannot know the real meaning of Vedanta-sutra, but Vyasadeva, who is an incarnation of Narayana, knows its real meaning. No ordinary living entity can interpret Vedanta-sutra according to his mundane conceptions. In order to curb commentaries on Vedanta-sutra by unscrupulous persons, the author himself, Vyasadeva, has already commentated upon the Vedanta-sutra by writing Srimad-Bhagavatam." In other words, the best explanation of a book is written by the author himself. No one can understand the author's mind unless the author himself discloses the purpose behind his writing. Therefore Vedanta-sutra should be understood through Srimad-Bhagavatam, the commentary written by the author of Vedanta-sutra.

Pranava, or omkara, is the divine substance of all the Vedas. Omkara is further explained in the Gayatri mantra exactly as it was explained in Srimad-Bhagavatam. There are four verses written in this connection, and these are explained to Brahma by Lord Krishna Himself. In his turn, Brahma explains them to Narada, and Narada explains them to Vyasadeva. In this way the purport of the verses of Srimad-Bhagavatam come down through disciplic succession. It is not that anyone and everyone can make his own foolish commentaries on Vedanta-sutra and mislead readers. Anyone who wants to understand Vedanta-sutra must read Srimad-Bhagavatam carefully. Under the instructions of Narada Muni, Vyasadeva compiled Srimad-Bhagavatam with the purpose of explaining the Vedanta-sutra. In writing Srimad-Bhagavatam, Vyasadeva collected all the essence of the Upanisads, the purpose of which was also explained in Vedanta-sutra. Srimad-Bhagavatam is thus the essence of all Vedic knowledge. That which is stated in the Upanishads and restated in Vedanta-sutra is explained very nicely in Srimad-Bhagavatam.

There is a passage in Isopanisad similar to a verse found in Srimad-Bhagavatam (8.1.10) which states that whatever one sees in the cosmic manifestation is but the Supreme Lord's energy and is nondifferent from Him. Consequently He is the controller, friend and maintainer of all living entities. We should live by the mercy of God and take only those things which are allotted to us. In this way, by not encroaching on another's property, one can enjoy life.

In other words, the purpose of the Upanishads, Vedanta-sutra and Srimad-Bhagavatam is one and the same. If one studies Srimad-Bhagavatam carefully, he will find that all the Upanishads and the Vedanta-sutra are nicely explained therein. Srimad-Bhagavatam teaches us how to re-establish our eternal relationship with the Supreme Lord, how to act in that relationship, and, lastly, how to achieve the highest benefit from it.

The four verses beginning with aham eva are the gist of the whole Bhagavatam. These are: "I am the supreme centre for the relationships of all living entities, and My knowledge is the supreme knowledge. That process by which I can be attained by the living entity is called abhidheya. By it, one can attain the highest perfection of life, love of Godhead. When one attains love of Godhead, his life becomes perfect." The explanation of these four verses is given in Srimad-Bhagavatam, and Lord Chaitanya gave a short description of the principles of these verses. He said that no one can understand the constitutional position of the Supreme Lord—His situation, His transcendental qualities, His transcendental activities and His six opulences. These cannot be understood by mental speculation or academic education; they can only be understood by the mercy of the Lord. As stated in Bhagwad Gita, one who is fortunate enough to receive the Lord's favour can understand all these explanations by the mercy of the Lord.

The Lord existed before the material creation; therefore the material ingredients, nature and the living entities all emanated from Him, and after dissolution they rest in Him. When the creation is manifest, it is maintained by Him; indeed, whatever manifestation we see is but a transformation of His external energy. When the Supreme Lord withdraws His external energy, everything enters into Him. In the first of the four verses, the word aham is given three times in order to stress that the

Supreme Personality of Godhead is full with all opulences. Aham is stated three times just to chastise one who cannot understand or believe in the transcendental nature and form of the Supreme Lord.

The Lord possesses His internal energy, His external, marginal and relative energies, the manifestation of the cosmic world and the living entities. The external energy is manifested by the qualitative modes (gunas) of material nature. One who can understand the nature of the living entity in the spiritual world can actually understand vedyam, or perfect knowledge. One cannot understand the Supreme Lord simply by seeing the material energy and the conditioned soul, but when one is in perfect knowledge, he is freed from the influence of the external energy. The moon reflects the light of the sun, and without the sun the moon cannot illuminate anything. Similarly, this material cosmic manifestation is but the refection of the spiritual world.

When one is actually liberated from the spell of the external energy, he can understand the constitutional nature of the Supreme Lord. Devotional service to the Lord is the only means by which the Lord can be attained, and this devotional service can be accepted by everyone and anyone in any country and under any circumstance. Devotional service is above the four principles of religion and the understanding derived from liberation. Even the preliminaries of devotional service are transcendental to the highest understanding derived from liberation and the understanding derived from ordinary religion.

One should therefore approach a bona fide spiritual master—irrespective of caste, creed, colour, country, etc.—and hear from him everything about devotional service. The real purpose of life is to revive our dormant love of God. Indeed, that is our ultimate necessity. How that love of God can be attained is explained in Srimad-Bhagavatam. There is theoretical knowledge and specific or realized knowledge, and perfect realized knowledge is attained when one realizes the teachings received from the spiritual master.

Personal and Impersonal Realization

The Puranas are called supplementary Vedic literatures. Because sometimes in the original Vedas the subject matter is too difficult for the common man to understand, the Puranas explain matters simply by the use of stories and historical incidents. In Srimad-Bhagavatam (10.14.32) it is stated that Maharaja Nanda and the cowherd men and inhabitants of Vrndavana are very fortunate because the Supreme Brahman, the Personality of Godhead, full of bliss, engages in His eternal pastimes as their friend.

According to Svetasvatara Upanishad, the apani-pado javano grahita mantra confirms that although Brahman has no material hands and legs, He nonetheless walks in a very stately way and accepts everything that is offered to Him. This suggests that He has transcendental limbs and is therefore not impersonal. One who does not understand the Vedic principles simply stresses the impersonal material features of the Supreme Absolute Truth and thus incorrectly calls the Absolute Truth impersonal.

The impersonalist Mayavadi philosophers want to establish the Absolute Truth as impersonal, but this is in contradiction to Vedic literature. Although Vedic literatures confirm the fact that the Supreme Absolute Truth has multiple energies, the Mayavadi impersonalists still try to establish that the Absolute Truth has no energy. The fact remains, however, that the Absolute Truth is full of energy and is a person as well. It is not possible to establish Him as impersonal.

According to the Vishnu Purana (6.7.61-3), the living entities are considered ksetrajna energy. Although the living entity is part and parcel of the Supreme Lord and is fully cognizant, he nonetheless becomes entrapped by material contamination and suffers all the miseries of material life. Such living entities live in different ways in accordance to the degree of their entanglement in material nature. The original energy of the Supreme Lord is spiritual and nondifferent from the Supreme Absolute Personality of Godhead. The living entity is called the marginal energy of the Supreme Lord, and the material energy is called the inferior energy. Due to his material inebriety, the living entity in the marginal position becomes entangled with the inferior energy, matter. At such a time he forgets his spiritual significance, identifies himself with material energy and thereby becomes subjected to the threefold miseries. Only when he is free from such material contamination can he be situated in his proper position.

According to Vedic instructions, one should understand the constitutional position of the living entity, the position of the Lord, and the position of material energy in their interrelation. First of all, one should try to understand the constitutional position of the Supreme Lord, the Personality of Godhead. That Supreme Lord has an eternal, cognizant, blissful body, and His spiritual energy is distributed as eternity, knowledge and bliss. In His blissful identity can be found His pleasure potency, and in His eternal identity He can be seen as the cause of everything. In His cognizant identity, He is the supreme knowledge. Indeed, the word Krishna indicates that supreme knowledge. In other words, the Supreme Personality, Krishna, is the reservoir of all knowledge, pleasure and eternity. The supreme knowledge of Krishna is exhibited in three different energies—internal, marginal and external. By virtue of His internal energy, He exists in Himself with His spiritual paraphernalia; by means of His marginal energy, He exhibits Himself as the living entities, and by means of His external energy He exhibits Himself as material energy. Behind each and every energetic exhibition there is the background of eternity, pleasure, potency and full cognizance.

The conditioned soul is the marginal potency overpowered by the external potency. However, when the marginal potency comes under the jurisdiction of the spiritual potency, it becomes eligible for love of Godhead. The Supreme Lord enjoys six kinds of opulences, and no one can establish that He is formless or that He is without energy. If someone claims so, his contention is completely opposed to the Vedic instructions. Actually the Supreme Personality of Godhead is the master of ali energies. It is only the living entity, who is an infinitesimal part and parcel of Him, who is overpowered by the material energy.

In the Mundaka Upanishad it is stated that there are two birds sitting on the same tree, and one of these birds is eating the fruit of this tree while the other bird is simply witnessing his activities. Only when the bird eating the fruit looks at the other bird does he become free from all anxieties. This is the position of the infinitesimal living entity. As long as he is forgetful of the Supreme Personality of Godhead, who witnesses all his activities, he is subjected to the threefold miseries. But when he looks to the Supreme Lord and becomes the Supreme Lord's devotee, he becomes free from all anxieties and material miseries.

The living entity is eternally subordinate to the Supreme Lord; the Supreme Lord is always the master of all energies, whereas the living entity is always under the domination of the Lord's energies. Although qualitatively one with the Supreme Lord, the living entity has the tendency to lord it over material nature; however, being infinitesimal, he is actually controlled by material nature. Thus the living entity is called the marginal potency of the Lord.

Because the living entity tends to be controlled by material nature, he cannot at any stage become one with the Supreme Lord. If a living entity were equal to the Supreme Lord, there would be no possibility of his being controlled by material energy. In Bhagwad Gita the living entity is described as one of the energies of the Supreme Lord. Although inseparable from the energetic, energy is still energy, and it cannot be equal with the energetic. In other words, the living entity is simultaneously one and different from the Supreme Lord. Bhagwad Gita (7.4-5) clearly states that earth, water, fire, air, ether, mind, intelligence and false ego are the eight elementary energies of the Supreme Lord and are of inferior quality, whereas the living entity is of superior quality. The Vedic literatures confirm the fact that the transcendental form of the Supreme Lord is eternal, blissful and full of knowledge.

The form of the Supreme Lord which is beyond the modes of material nature is not like the forms of this material world. His form is fully spiritual and cannot be compared with any material form. According to Vedic literatures, one who does not accept the spiritual form of the Supreme Lord is an atheist. Because Lord Buddha did not accept these Vedic principles, the Vedic teachers consider him to be an atheist. Although Mayavadi philosophers pretend to accept the Vedic principles, they indirectly preach Buddhist philosophy, or atheistic philosophy, and do not accept the Supreme Personality of Godhead. Mayavadi philosophy is inferior to Buddhist philosophy, which directly denies Vedic authority. Because it is disguised as Vedanta philosophy, Mayavadi philosophy is more dangerous than Buddhism or atheism.

Vedanta-sutra is compiled by Vyasadeva for the benefit of all living entities. It is through Vedanta-sutra that the philosophy of bhakti-yoga can be understood. Unfortunately, the Mayavadi commentary, Sariraka-bhasya, has practically defeated the purpose of Vedanta-sutra. In the Mayavadi commentary, the spiritual, transcendental form of the Supreme Personality of Godhead has been denied, and the Supreme Brahman has been dragged down to the level of the individual Brahman, the living entity. Both the Supreme Brahman and the individual Brahman have been denied spiritual form and individuality, although it is clearly stated that the Supreme Lord is the one supreme living entity and the other living entities are the many subordinate living entities.

Thus reading the Mayavadi commentaries on Vedanta-sutra is always dangerous. The chief danger is that through these commentaries one may come to consider the living entity to be equal to the Supreme Lord. It is easy for a conditioned living entity to be falsely directed in this way, and once he is so directed he can never come to his actual position or enjoy his eternal activity in bhakti yoga. In other words, the Mayavadi philosophy has rendered the greatest disservice to humanity by promoting the impersonal view of the Supreme Lord. Thus Mayavadi philosophers deprive human society of the real message of Vedanta-sutra.

From the very beginning of Vedanta-sutra it is accepted that the cosmic manifestation is but an energetic display of the Supreme Lord. The very first aphorism (janmady asya) describes the Supreme Brahman as He from whom everything emanates.

Everything is maintained by Him, and everything is dissolved in Him. Thus the Absolute Truth is the cause of creation, maintenance and dissolution. The cause of a piece of fruit is the tree; when a tree produces a piece of fruit, one cannot say that the tree is impersonal. The tree may produce hundreds and thousands of fruits, but it remains as it is. The fruit is produced, and it develops and stays for some time; then it dwindles and vanishes.

This does not mean that the tree also vanishes. Thus from the very beginning the Vedanta-sutra explains the doctrine of by-products. These activities of production, maintenance and dissolution are carried out by the inconceivable energy of the Supreme Lord. The cosmic manifestation is a transformation of the energy of the Supreme Lord, although the energy of the Supreme Lord and the Supreme Lord Himself are nondifferent and inseparable. A touchstone may produce great quantities of gold in contact with iron, but still the touchstone remains as it is. Despite His producing huge material cosmic manifestations, the Supreme Lord is always in His transcendental form.

Mayavadi philosophy has the audacity to reject the purpose of Vyasadeva, as explained in the Vedanta-sutra, and to attempt to establish a doctrine of transformation which is totally imaginary. According to the Mayavadi philosophy, the cosmic manifestation is but the transformation of the Absolute Truth, and the Absolute Truth has no separate existence outside the cosmic manifestation. This is not the message of Vedanta-sutra. The transformation has been explained by Mayavadi philosophers as false, but it is not false. It is only temporary. The Mayavadi philosophers maintain that the Absolute Truth is the only truth and that this material manifestation known as the world is false. Actually this is not the case. The material contamination is not exactly false; because it is relative truth, it is temporary. There is a difference between something that is temporary and something that is false.

Pranava, or omkara, is the chief vibration found in the Vedic hymns, and omkara is considered to be the sound form of the Supreme Lord. From omkara all Vedic hymns have emanated, and the world itself has also emanated from this omkara sound. The words tat tvam asi, also found in the Vedic hymns, are not the chief vibrations but are explanations of the constitutional position of the living entity. Tat tvam asi means that the living entity is a spiritual particle of the supreme spirit, but this is not the chief motif of the Vedanta or Vedic literatures. The chief sound representation of the Supreme is omkara.

All these faulty explanations of Vedanta-sutra are considered atheistic. Because the Mayavadi philosophers do not accept the eternal transcendental form of the Supreme Lord, they are unable to engage in real devotional service. Thus the Mayavadi philosopher is forever bereft of Krishna consciousness and Krishna's devotional service. The pure devotee of the Personality of Godhead never accepts the Mayavadi philosophy as an actual path to transcendental realization. The Mayavadi philosophers hover in the moral and immoral material atmosphere of the cosmic world and consequently are always engaged in rejecting and accepting material enjoyment. They have falsely accepted the nonspiritual as the spiritual, and as a result they have forgotten the spiritual eternal form of the Supreme Personality of Godhead, as well as His name, quality and entourage. They consider the transcendental pastimes, name, form and qualities of the Supreme to be products of material nature. Because of their acceptance and rejection of material pleasure and misery, the Mayavadi philosophers are eternally subjected to material misery.

The actual devotees of the Lord are always in disagreement with the Mayavadi philosophers. There is no way that impersonalism can possibly represent eternity, bliss and knowledge. Being situated in imperfect knowledge of liberation, the Mayavadi decries eternity, knowledge and bliss as materialism. Because they reject devotional service, they are unintelligent and unable to understand the effects of devotional service. The word jugglery they use in an attempt to amalgamate knowledge, the knowable and the knower simply reveals them to be unintelligent. The doctrine of by-product is the real purport of the beginning of Vedanta-sutra. The Lord is empowered with innumerable unlimited energies, and consequently He displays the by-products of these energies in different ways. Everything is under His control. The Supreme Lord is also the supreme controller, and He is manifested in innumerable energies and expansions.

Lord Chaitanya and Ramananda Raya

The author of Chaitanya-caritamrta has described Lord Chaitanya Mahaprabhu as the ocean of transcendental knowledge and Sri Ramananda Raya as the cloud which is produced from that ocean. Ramananda Raya was a greatly advanced scholar in devotional service, and by the grace of Lord Chaitanya he gathered all transcendental conclusions just as a cloud gathers water from the ocean. As clouds appear from the ocean, distribute water all over the world, and return to the ocean, so by the grace of Lord Chaitanya, Ramananda Raya attained his higher knowledge of devotional service and again, after retiring from service, resolved to see Lord Chaitanya in Puri.

When Lord Chaitanya visited the southern part of India, He first went to the great temple known as Jiyara-Narsimha-ksetra. This temple is situated in a place known as Simhacalam, five miles from the Vishakhapattanam railway station. The temple is situated on the top of a hill. There are many temples in that area, but Jiyara-Narsimha-ksetra temple is the largest of all. This temple is filled with beautiful sculpture, of interest to many students, and due to its popularity it is a very rich temple. An inscription in the temple states that the King of Vijayanagar formerly decorated this temple with gold and even smeared the body of the Deity with gold. To facilitate attendance, there are free apartments for visitors at the temple. The temple is managed by priests of the Ramanujacharya sect.

When Lord Chaitanya visited this temple, He praised the Deity and quoted a verse from Sridhara Swami's commentary on Srimad-Bhagavatam (7.9.1):

ugro 'py anugra evayam
sva-bhaktanam nrkesari

kesariva svapotanam
anyesam ugra-vikramah

"Although Lord Narsimha is very severe to demons and nondevotees, He is very kind to His submissive devotees like Prahalad." Lord Narsimha appeared as a half-man, half-lion incarnation of Krishna when Prahalad, a boy devotee of the Lord, was harassed by his demoniac father Hiranyakasipu. Just as a lion is very ferocious to other animals but very kind and submissive to his cubs, so Lord Narsimha appeared ferocious to Hiranyakasipu and very kind to His devotee Prahalad.

After visiting the temple of Jiyara-Narsimha, the Lord proceeded further south into India and ultimately reached the bank of the Godavari. While on the bank of this river, the Lord remembered the Yamuna River in Vrndavana, and He considered the trees on the bank to be the forest of Vrndavana. Thus He was in ecstasy there. After taking a bath on the banks of the Godavari, the Lord sat near the bank and began chanting Hare Krishna, Hare Krishna, Krishna Krishna, Hare Hare. Hare Rama, Hare Rama, Rama Rama, Hare Hare.

While sitting and chanting, the Lord saw that the governor of the province, Sri Ramananda Raya, had reached the banks of the river accompanied by his associates, which included many brahmanas. Previously the Lord had been asked by Sarvabhauma Bhattacharya to visit the great devotee Ramananda Raya at Kabur. The Lord could understand that the man approaching the river bank was Ramananda Raya, and He desired to see him immediately. However, because He was in the renounced order of life, He restrained Himself from going to see a political personage. Being a great devotee, Ramananda Raya was attracted by the features of Lord Chaitanya, who appeared as a sanyasi, and he himself came to see the Lord. Upon reaching Chaitanya Mahaprabhu, Ramananda Raya prostrated himself and offered his obeisances and respects. Lord Chaitanya received him by vibrating Hare Krishna, Hare Krishna, Krishna Krishna, Hare Hare. Hare Rama, Hare Rama, Rama Rama Hare Hare.

When Ramananda Raya presented his credentials, Lord Chaitanya embraced him, and both of them were overwhelmed with ecstasy. The brahmanas who accompanied Ramananda Raya were surprised to see them embracing in transcendental ecstasy. The brahmanas were all stalwart followers of the rituals, and they could not understand the meaning of such devotional symptoms. Indeed, they were rather surprised to see such a great sanyasi touch a sudra, and they were also surprised to see Ramananda Raya, who was a great governor and practically king of that province, crying simply by touching a sanyasi. Lord Chaitanya understood the brahmanas' thoughts, and, considering the unfavorable situation, He pacified Himself.

After this, Lord Chaitanya and Ramananda Raya sat down together. "Sarvabhauma Bhattacharya has spoken very highly of you," Lord Chaitanya informed him. "So I have come to see you."

"Sarvabhauma Bhattacharya considers me to be one of his devotees," Ramananda Raya replied. "Therefore he has kindly recommended that You see me."

Ramananda Raya very much appreciated the Lord's touching a man of wealth. A king, governor or any politician is always absorbed in thoughts of political affairs and pounds-shilling-pence; therefore such persons are avoided by sannyasis. Lord Chaitanya, however, knew Ramananda Raya to be a great devotee, and therefore He did not hesitate to touch and embrace him. Ramananda Raya was surprised by Lord Chaitanya's behaviour, and he cited a verse from Srimad-Bhagavatam (10.8.4): "The great personalities and sages appear in the homes of worldly men just to show them mercy."

Lord Chaitanya's special treatment of Ramananda Raya indicated that although Ramananda Raya was born in a nonbrahminical family, he was far, far advanced in spiritual knowledge and activity. Therefore he was more respectable than one who simply happens to be born in a brahminical family. Although Ramananda, out of his meek and gentle behaviour, considered himself to be born in a lower sudra family, Lord Chaitanya nonetheless considered him to be situated in the highest transcendental stage of devotion. Devotees never advertise themselves as great, but the Lord is very anxious to advertise the glory of His devotees. After meeting for the first time that morning on the banks of the Godavari, Ramananda Raya and Lord Chaitanya separated with the understanding that Ramananda Raya would come in the evening to see the Lord.

That evening, after the Lord had taken His bath and seated Himself, Ramananda Raya came to see Him with a servant. He offered his respects and sat down before the Lord. Before Ramananda Raya could even ask the Lord a question about the advancement of spiritual knowledge, the Lord Himself said, "Please quote some verses from scripture about the ultimate goal of human life."

Sri Ramananda Raya at once replied: "A person who is sincere in his occupational duty will gradually develop a sense of God consciousness." He also quoted a verse from Vishnu Purana (3.8.9) which states that the Supreme Lord is worshipped by one's occupational duty and that there is no alternative for satisfying Him. The purport is that human life is meant for understanding one's relationship with the Supreme Lord, and by acting in that way any human being can dovetail himself in the service of the Lord by discharging his prescribed duties. For this purpose human society is divided into four classes: the intellectuals (brahmanas), the administrators (ksatriyas), the merchants (Vaishyas), and the labourers (sudras). For each class there are prescribed rules and regulations as well as occupational functions. The prescribed duties and qualities of the four classes are described in Bhagwad Gita (18.41-44). A society which is civilized and organized should follow the prescribed rules and regulations for the particular classes. At the same time, for spiritual advancement, the four stages of asrama must also be followed: namely, student life (brahmacarya), householder (grhastha), retired (vanaprastha) and the renounced life (sannyasa).

Ramananda Raya stated that those who strictly follow the rules and regulations of these eight social divisions can actually satisfy the Supreme Lord, and one who does not follow them certainly spoils his human form of life and glides toward hell. One can peacefully execute the goal of human life simply by following the rules and regulations which apply to one's self.

The character of a particular person develops by following the regulative principles in accordance with one's birth, association and education.

The divisions of society are so designed that many people of different character can be regulated under them for the peaceful administration of society and for spiritual advancement as well. The social classes can be further characterized as follows: (1) He whose aim is to understand the Supreme Lord, the Personality of Godhead, and devote himself to the learning of the Vedas and similar literatures is called a brahmana. (2) He who has taken to displaying force and entering government administration is called a Kshatriya. (3) He who is engaged in agriculture, herding cows and carrying out a trade or business is called a vaisya. (4) He who has no special knowledge but is satisfied by serving the other three classes is called a sudra. If one faithfully discharges his prescribed duties, he is sure to advance toward perfection.

Thus regulated life is the source of perfection for everyone. When regulated life culminates in devotional service to the Lord, one attains his perfection. Otherwise such regulations are simply a useless waste of time.

After hearing Ramananda Raya expound upon the proper execution of a regulated life, Lord Chaitanya said that such regulations are simply external. Indirectly He asked Ramananda to expound on something superior to such an external exhibition. Formal execution of rituals and religion is useless unless it culminates in the perfection of devotional service. Lord Vishnu is not satisfied simply by a ritualistic adherence to Vedic instructions; He is actually pleased when one attains the stage of devotional service.

According to the verse cited by Ramananda Raya, one can rise to the point of devotional service by ritualistic performance. In Bhagwad Gita, Sri Krishna, who appeared to deliver all classes of people, states that a human being can attain the highest perfectional stage of life by worshipping the Supreme Lord, from whom everything has emanated, through his occupational duty.

sve sve karmany abhiratah
samsiddhim labhate narah
svakarma-niratah siddhim
yatha vindati tac chrnu

yatah pravrttir bhutanam
yena sarvam idam tatam
svakarmana tam abhyarcya
siddhim vindati manavah

"By following his qualities of work, every man can become perfect. Now please hear from Me how this can be done. By worship of the Lord, who is the source of all beings and who is all-pervading, man can, in the performance of his own duty, attain perfection." (Bg. 18.45-46) This perfectional process is followed by great devotees like Bodhayana, Tanka, Dramida, Guhadeva, Kapardi and Bharuci. All these great personalities have followed this particular path of perfection.

The Vedic injunctions also aim in this direction. Ramananda Raya wanted to present these facts before the Lord, but apparently discharge of ritualistic duties wasn't sufficient, for Lord Chaitanya said that it was external. Lord Chaitanya was pointing out that if a man has a material conception of life, he cannot attain the highest perfection even if he follows all the ritualistic regulations.

Relationship with the Supreme

Lord Chaitanya rejected the statement cited by Ramananda Raya from Vishnu Purana because the Lord wished to reject a class of philosophers known as karma-mimamsa. Karma-mimamsa followers accept God to be subject to one's work. Their conclusion is that if one works nicely, God is bound to give good results. Thus one can understand from the statement of Vishnu Purana that Vishnu, the Supreme Lord, has no independence but is bound to award a certain kind of result to the worker. Such a dependent goal becomes subjected to the worshipper, who accepts the Supreme Lord to be both impersonal and personal, as he may wish. Actually this philosophy stresses the impersonal feature of the Supreme

Absolute Truth. Because Lord Chaitanya did not like such impersonalism, He rejected it.

"Tell Me if you know something beyond this conception of the Supreme Absolute Truth," Lord Chaitanya finally said.

Ramananda Raya understood the purpose of Lord Chaitanya, and, stating that it is better to give up the results of fruitive activities, he quoted a verse from Bhagwad Gita:

yat karosi yad asnasi
yaj juhosi dadasi yat
yat tapasyasi kaunteya
tat kurusva mad-arpanam

"O son of Kunti, all that you do, all that you eat, all that you offer and give away, as well as all austerities that you may perform, should be done as an offering unto Me.". (Bg. 9.27) There is also a similar passage in Srimad-Bhagavatam (11.2.36) which states that one should submit everything—his fruitive activities, body, speech, mind, senses, intelligence, soul and modes of nature—to the Supreme Personality of Godhead, Narayana.

Lord Chaitanya, however, also rejected this second statement, saying, "If you know of something higher, state it."

Offering everything to the Supreme Personality of Godhead, as enjoined by Bhagwad Gita and Srimad-Bhagavatam, is better than impersonally making the Supreme Lord subject to our work, but it is still short of surrendering activities to the Supreme Lord. A worker's identification with material existence cannot be changed without proper guidance. Such fruitive activity will continue one's material existence. A worker is simply instructed here to offer the results of his work to the Supreme Lord, but there is no information given to enable one to get out of the material entanglement. Therefore Lord Chaitanya rejected his proposal.

After having his suggestions rejected twice, Ramananda proposed that one should forsake his occupational activities altogether and by detachment rise to the transcendental plane. In other words, he recommended complete renunciation of worldly life, and to support this view he cited evidence from

Srimad-Bhagavatam (11.11.32) wherein the Lord says: "In the scriptures I have described the ritualistic principles and the way one can become situated in devotional service. That is the highest perfection of religion." Ramananda also quoted Lord Krishna's injunction in Bhagwad Gita:

sarva-dharman parityajya
mam ekam saranam vraja
aham tvam sarva-papebhyo
mokshayisyami ma sucah

"Abandon all varieties of religion and just surrender unto Me. I shall deliver you from all sinful reactions. Do not fear." (Bg. 18.66)

Lord Chaitanya also rejected this third proposal, for He wanted to demonstrate that renunciation in itself is not sufficient. There must be positive engagement. Without positive engagement, the highest perfectional stage cannot be attained. Generally there are two kinds of philosophers in the renounced order of life. The goal of one is nirvana, and the goal of the other is the impersonal Brahman effulgence. Such philosophers cannot imagine that they can reach beyond nirvana and the Brahman effulgence to the Vaikuntha planets of the spiritual sky. Because in simple renunciation there is no conception of spiritual planets and spiritual activities, Lord Chaitanya rejected this third proposal.

Ramananda Raya then cited more evidence from Bhagwad Gita:

brahma-bhutah prasannatma
na socati na kanksati
samah sarvesu bhutesu
mad-bhaktim labhate param

"One who is thus transcendentally situated at once realizes the Supreme Brahman. He never laments, nor desires to have anything; he is equally disposed to every living entity. In that state he attains pure devotional service unto Me." (Bg. 18.54) Ramananda Raya first suggested devotional service rendered with renunciation of fruitive activities, but here he suggests that devotional service with full knowledge and spiritual realization added is superior. Lord Chaitanya, however, rejected this proposal also because simply by renouncing material results in Brahman realization one does not realize the spiritual world and spiritual activities. Although there is no material contamination when one attains the stage of Brahman realization, that stage is not perfect because there is no positive engagement in spiritual activity. Because it is still on the mental plane, it is external. The pure living entity is not liberated unless he is completely engaged in spiritual activity.

As long as one is absorbed in impersonal thoughts or in thoughts of the void, his entrance into an eternal blissful life of knowledge is not completed. When spiritual knowledge is not complete, one will be hindered in his attempt to cleanse the mind of all material variegatedness. Thus impersonalists are frustrated in their attempts to make the mind void by artificial meditation. It is very difficult to void the mind of all material conceptions. As stated in Bhagwad Gita:

kleso 'dhikataras tesam
avyaktasakta-cetasam
avyakta hi gatir duhkham
dehavadbhir avapyate

"For those whose minds are attracted to the unmanifested, impersonal feature of the Supreme, advancement is very troublesome. To make progress in that discipline is always difficult for those who are embodied." (Bg. 12.5) The liberation which is obtained by such impersonal meditation is not complete; therefore Lord Chaitanya rejected it.

After his fourth proposal was rejected, Ramananda Raya said that devotional service rendered without any attempt at cultivation of knowledge or mental speculation is the highest stage of perfection. To support this view, he gave evidence from Srimad-Bhagavatam (10.14.3) wherein Lord Brahma tells the Supreme Personality of Godhead:

jnane prayasam udapasya namanta eva
jivanti san-mukharitam bhavadiya-vartam
sthane sthitah sruti-gatam tanu-van-manobhir
ye prayaso 'jita jito 'py asi tais tri-lokyam

"My dear Lord, one should give up monistic speculation and the cultivation of knowledge altogether. He should begin his spiritual life in devotional service by receiving information of the Lord's activities from a realized devotee of the Lord. If one cultivates his spiritual life by following these principles and keeping himself on the honest path in life, then although Your Lordship is never conquered, You become conquered by such a process."

When Ramananda Raya presented this proposal, Lord Chaitanya at once said, "Yes, this is right." In this age there is no possibility of acquiring spiritual knowledge by renunciation, by mixed devotional service, by fruitive activity in mixed devotional service, or by the culture of knowledge. Because most people are fallen and because there is no time to elevate them by a gradual process, the best course, according to Lord Chaitanya, is to let them remain in whatever condition they are in but to engage them in hearing of the activities of the Supreme Lord as those activities are explained in Bhagwad Gita and Srimad-Bhagavatam.

The transcendental messages of the scriptures should be received from the lips of realized souls. In this way a person may continue to live in whatever condition he is in and still make progress in spiritual advancement. Thus one can surely advance and fully realize the Supreme Personality of Godhead. Although Lord Chaitanya accepted these principles, He still requested Ramananda Raya to further explain advanced devotional service. Thus Lord Chaitanya gave Ramananda Raya a chance to discuss gradual advancement from the principles of varnasrama-dharma (the four castes and four orders of spiritual life). Lord Chaitanya rejected the varnasrama-dharma and the offering of fruitive activity because in the field of executing pure devotional service, there is very little use for such principles.

Without self-realization, the artificial methods of devotional service cannot be accepted as pure devotional service. Self-realized pure devotional service is completely different from all other kinds of transcendental activity. The highest stage of transcendental activity is always free from all material desires, fruitive efforts and speculative attempts at knowledge. The highest stage concentrates on the simple, favourable execution of pure devotional service.

Ramananda Raya could understand the motive of Lord Chaitanya; therefore he stated that attainment of pure love of Godhead is the highest perfectional stage. There is a very nice verse in Padyavali which is said to be composed by Ramananda Raya himself. The purport of the verse is: "As long as there is hunger in the belly and one feels like eating and drinking, one can become happy by taking anything eatable. Similarly, there may be much paraphernalia for worshipping the Supreme Lord, but when that is mixed with pure love of Godhead, it becomes an actual source of transcendental happiness." Ramananda Raya also composed another verse which stated that even after millions and millions of births one cannot achieve a sense of devotional service, but if, somehow or other, one desires to attain devotional service, the association of a pure devotee will render it possible.

Thus one should have a strong desire to engage in devotional service. In these two verses, Ramananda Raya has described the regulative principles and developed love of Godhead. Lord Chaitanya wanted to bring him to the stage of developed love of Godhead, and He wanted him to speak from that platform. Thus the discussion between Ramananda Raya and Lord Chaitanya proceeds on the basis of love of Godhead. If love of Godhead is elevated to the personal platform, it is called prema-bhakti.

In the beginning of prema-bhakti, a particular relationship between the Supreme Lord and the devotee is not established, but when prema-bhakti develops, a relationship with the Supreme Lord is manifested in different transcendental flavours. The first stage is that of servitude, wherein the Supreme Lord is accepted as the master and the devotee as the eternal servitor. When Lord Chaitanya accepted this process, Ramananda Raya described the relationship between the servitor and the master. As described in Srimad-Bhagavatam (9.5.16), Durvasa Muni, a great mystic yogi who considered himself very elevated, envied Maharaja Ambarisa, who was known as the greatest devotee of the time. In an attempt to harass Maharaja Ambarisa, Durvasa Muni met with a great catastrophe and was defeated by the sudarsana-cakra of the Lord. Durvasa Muni admitted his fault and said, "For pure devotees who are always engaged in the transcendental loving service of the Lord, nothing is considered impossible, for they are engaged in the service of the Supreme Lord, whose very name is sufficient for liberation."

In Stotra-ratna (46), Yamunacarya writes: "My Lord, those who keep themselves independent of Your service are helpless. They work on their own account, and they receive no support from superior authority. Therefore I long for the time when I shall engage fully in Your transcendental loving service without any desire for material satisfaction and without being confined to the mental plane. Only when I engage in such unalloyed devotional service will I enjoy actual spiritual life."

Upon hearing this statement, the Lord requested Ramananda Raya to go even further.

Pure Love for Krishna

Encouraged by Lord Chaitanya to proceed further, Ramananda Raya said that the fraternal relationship with Lord Krishna is on an even higher transcendental plane. In this way Ramananda Raya points out that when the relationship with Krishna increases in affection, the mood of fear and the consciousness of the superiority of the Supreme Lord diminish. At this point, the mood of faith increases, and this faith is called friendship. In the fraternal relationship, there is a sense of equality between Krishna and His friends.

In this regard, Ramananda Raya quoted a verse from Srimad-Bhagavatam (10.12.11) in which Sukadeva Gosvami describes Lord Krishṇa's lunch with His friends in the forest. Lord Krishna and His friends went to the forest to play with the cows, and it is said that the boys who accompanied Krishna enjoyed transcendental friendship with the Supreme Personality of Godhead. The Supreme Lord is considered to be the impersonal Brahman by great sages, the Supreme Personality of Godhead by the devotees, and an ordinary human being by common men.

Lord Chaitanya appreciated this statement very much, yet He said, "You can go even further." Being so requested, Ramananda Raya then stated that the paternal relationship with Krishna is a still higher transcendental position. When the affection of the fraternal relationship increases, it develops into the paternal relationship that is found between father and son. Regarding this, Ramananda Raya quoted a verse from Srimad-Bhagavatam (10.8.46) wherein Maharaja Pariksit inquired from Sukadeva Gosvami about the magnitude of righteous activity performed by Yashoda, the mother of Krishna, enabling her to be called "mother" and her breasts be sucked by the Supreme Personality of Godhead. He also quoted another verse from Srimad-Bhagavatam (10.9.20) in which it is stated that Yashoda received such incomparable mercy from the Supreme Personality of Godhead that it cannot even be compared to the mercy received by Brahma, the first created living being, or by Lord Shiva, or even by the goddess of fortune, Lakshmi, who is always situated on the chest of Lord Vishnu.

Lord Chaitanya then asked Ramananda Raya to proceed further in order to come to the point of conjugal love. Understanding the mind of Lord Chaitanya, Ramananda Raya immediately answered that it was indeed conjugal love with Krishna that constituted the highest relationship. In other words, intimate relationships with Krishna develop from an ordinary conception of the Supreme Personality of Godhead, to the conception of master and servant, and, when this becomes confidential, it develops into a friendly relationship, and when this relationship further develops, it becomes paternal, and when this develops to the highest point of love and affection, it is known as conjugal love with the Supreme Lord.

Ramananda Raya quoted another verse from Srimad-Bhagavatam (10.47.60) stating that the transcendental mode of ecstasy exhibited during the rasa dance between the gopis and Krishna was never relished even by the goddess of fortune, who is always situated on the chest of the Lord in the spiritual kingdom. And what to speak of the experience of ordinary women?

Ramananda Raya then explained the gradual process by which pure love for Krishna is developed. He pointed out that the living entity is related to the Supreme Personality of Godhead in one of the relationships just suitable for him. Actually, relationships with the Supreme Lord begin with the master and servant relationship and further develop into friendship, paternal love and conjugal love. One who attains his particular relationship with the Supreme Personality of Godhead should be known to be in the best relationship for him, but when these transcendental relationships are studied, it can be seen that the neutral stage of realization (brahma-bhuta) is the first. When one accepts the Lord as master and himself as servant, the relationship develops, and it develops further when one becomes a friend of the Supreme Lord, and yet still further when one becomes a father. Thus the relationship advances from friendship to paternal love and finally to conjugal love, which is the supreme relationship with the Lord.

Self-realization in the relation as servitor is certainly transcendental, and when a sense of fraternity is added, the relationship develops. As affection increases, this relationship develops into paternity and conjugal love. Ramananda Raya quoted a verse from Bhakti-rasamrta-sindhu (2.5.38) stating that spiritual affection for the Supreme Lord is transcendental in all cases, but the individual devotee has a specific aptitude for a particular relationship, and that relationship is more relishable for him than the others.

Such transcendental relationships with the Supreme Lord cannot be manufactured by the mental concoctions of pseudo-devotees. In Bhakti-rasamrta-sindhu (1.2.101) Rupa Gosvami has stated that devotional service that makes no reference to Vedic scriptures or Vedic literatures and that does not follow the principles set forth therein can never be approved. Sri Bhaktisiddhanta Saraswati Gosvami Maharaja has also remarked that professional spiritual masters, professional Bhagavatam reciters, professional kirtana performers and those engaged in self-concocted devotional service cannot be accepted.

In India there are various professional communities known as Aula, Vaula, Kartabhaja, Neda Daravesa, Snai, Atibadi,

Cudadhari, and Gauranganagari. A member of the Ventor Gosvami Society, or the caste called Gosvami, cannot be accepted as a descendant of the six original Gosvamis. Nor can so-called devotees who manufacture songs about Lord Chaitanya, nor those who are professional priests or paid reciters, be accepted. One who does not follow the principles of pancaratra, or one who is an impersonalist or addicted to sex life, cannot be compared with those who have dedicated their lives to the service of Krishna. A pure devotee who is always engaged in Krishna consciousness can sacrifice everything for the service of the Lord. One who has dedicated his life to the service of Lord Chaitanya, Krishna and the spiritual master, or a person who is following the principles of householder life, as well as one following the principles of the renounced life in the order of Chaitanya Mahaprabhu, is a devotee and cannot be compared to professional men.

When one is freed from all material contaminations, any one of the relationships with Krishna is transcendentally relishable. Unfortunately, those who are inexperienced in the transcendental science cannot appreciate the different relationships with the Supreme Lord. They think that ali such relationships arise from maya. Chaitanya-caritamrta states that earth, water, fire, air and ether (the five gross elements) are developed from subtle forms to grosser forms. For example, sound is found in ether, but in air there is sound and touch. When fire is added, there is sound, touch and form as well. When water is added, there is sound, touch, form and taste, and when earth is added, there is sound, touch, form, taste and smell. Just as various characteristics develop in the progression from ether down to earth, so the five characteristics of devotion develop and are all found in the relationship of conjugal love. Thus the relationship with Krishna in conjugal love is accepted as the highest perfectional stage of love of God.

As stated in Srimad-Bhagavatam (10.82.44): "Devotional service to the Supreme Personality of Godhead is the life of every living entity." Indeed, the Lord informed the damsels of Vraja that their love for Him was the only cause of their achieving His association. It is said that Lord Krishna, in relation with His devotees, accepts all kinds of devotional service according to the aptitude of the devotee. In this way, Krishna reciprocates to a devotee's needs. If one wants a relationship with Krishna as master and servant, Krishna plays the part of the perfect master. For one who wants Krishna as a son in the parental relationship, Krishna plays the part of a perfect son. Similarly, if a devotee wants to worship Krishna in conjugal love, Krishna plays the part of a husband or paramour perfectly. However, Krishna Himself has admitted that His loving relationship with the damsels of Vraja in conjugal love is the highest perfectional stage. In Srimad-Bhagavatam (10.32.22) Krishna told the gopis:

na paraye 'ham niravadya-samyujam
sva-sadhu-krtyam vibudhayusapi vah
ya mabhajan durjaya-geha-srnkhalah
samvrscya tad vah pratiyatu sadhuna

"Your relationship with Me is completely transcendental, and it is not possible for Me to offer anything in exchange for your love, even after many births. You have been able to give up all attachment for material enjoyment, and you have searched after Me. Since I am unable to repay your love, you have to be pleased with your own activities."

Srila Bhaktisiddhanta Saraswati Gosvami Maharaja has remarked that there is a class of common men who claim that anyone and everyone can worship the Supreme Lord according to his own invented mode of worship and still attain the Supreme Personality of Godhead. They claim that one can approach the Supreme Lord either through fruitive activities, speculative knowledge, meditation or austerity and that any one of the methods will succeed. They claim that one can accept many different paths and still reach the same place, and they maintain that the Supreme Absolute Truth may be worshipped either as the Goddess Kali, or Goddess Durga, or Lord Shiva, Ganesha, Rama, Hari, or Brahma. In short, they maintain that it does not matter how the Absolute Truth is addressed, for ali names are one and the same. They give the example of a man with many names; if he is called by any of those names, he will answer.

Such views may be very pleasing to an ordinary person, but they are full of misconceptions. One who worships the demigods, motivated by material lust, cannot attain the Supreme Personality of Godhead. If one worships the demigods, the external energy of the Lord can award some results, but this is not to say that one can attain the Supreme Lord by such worship. Indeed, their worship is discouraged in Bhagwad Gita:

antavat tu phalam tesam
tad bhavaty alpa-medhasam
devan deva-yajo yanti
mad-bhakta yanti mam api

"Men of small intelligence worship the demigods, and their fruits are limited and temporary. Those who worship the demigods go to the planets of the demigods, but My devotees ultimately reach My supreme planet." (Bg. 7.23) Thus the Supreme Lord awards the benediction of His association only to those who worship Him, and not to those who worship the demigods. It is not a fact that everyone and anyone can reach the Supreme Personality of Godhead by worshipping material demigods.

It is therefore surprising that a man can imagine that he will become perfect by worshipping the demigods. The results of devotional service rendered in full Krishna consciousness cannot be compared to the results of demigod worship, fruitive activity or mental speculation. By the results of fruitive activities, one can either go to the heavenly planets or the hellish ones.

The Transcendental Pastimes of Radha and Krishna

The difference between executing ordinary religious activities and devotional service is very great. By executing religious rituals one can achieve economic development, sense gratification or liberation (merging into the existence of the Supreme), but the results of transcendental devotional service are completely different from such temporary benefits. Devotional service of the Lord is ever green, and it is increasingly transcendentally pleasing. Thus there is a gulf of difference between the results derived from devotional service and those derived from religious rituals. The great spiritual energy known as jadadhisthatri, or mahamaya, the superintendent of the material world, and the material departmental directors, the demigods, as well as the products of the external energy of the Supreme Lord, are but perverted reflections of the opulence of the Supreme Lord.

The demigods are actually order carriers of the Supreme Lord, and they help manage the material creation. In Brahma-samhita it is stated that the workings of the supremely powerful superintendent, Durga, are but shadowy indications of the workings of the Supreme Lord. The sun works just like the eye of the Supreme Lord, and Brahma works just as the reflected light of the Supreme Lord. Thus all the demigods as well as the external energy herself, Durgadevi, and all the different departmental directors are but servants of the Supreme Lord in the material world.

In the spiritual world, there is another energy, the superior spiritual energy, or internal energy, which acts under the direction of yogamaya. Yogamaya is the internal potency of the Supreme Lord; she also works under the Lord's direction, but she works in the spiritual world. When the living entity puts himself under the direction of yogamaya instead of mahamaya, he gradually becomes a devotee of Krishna. Yet those who are after material opulence and material happiness place themselves under the care of the material energy, mahamaya, or under the care of material demigods like Lord Shiva and others. In Srimad-Bhagavatam it is found that when the gopis of Vrndavana desired Krishna as their husband, they prayed to the spiritual energy, yogamaya, for the fulfilment of their desire. In the Sapta-sati it is found that King Suratha and a merchant named Samadhi worshipped mahamaya for material opulence. Thus one should not mistakenly equalize yogamaya an d mahamaya.

Because the Lord is on the absolute platform, there is no difference between the holy name of the Lord and the Supreme Lord Himself. There are different names for the Supreme Lord, and these names have different purposes and meanings. For instance, He is known as Paramatma, the Supersoul, Brahman the Supreme Absolute, Srstikarta the creator, Narayana the transcendental Lord, Rukminiramana the husband of Rukmini, Gopinatha the enjoyer of the gopis, and Krishna. In this way the Lord has different names, and these names indicate different functions.

The aspect of the Supreme Lord as the creator is different from the aspect of the Lord as Narayana. Some of the names of the Lord as the creator are conceived by materialistic men. One cannot fully realize the essence of the Supreme Personality of Godhead by understanding the name of the creator because this material creation is a function of the external energy of the Supreme Lord. Thus the conception of God as the creator includes only the external feature. Similarly, when we call the Supreme Lord Brahman, we cannot have any understanding of the six opulences of the Supreme Lord. In Brahman realization, the six opulences are not realized in full; therefore Brahman realization is not complete understanding of the Supreme Lord. Neither is Paramatma realization, realization of the Supersoul, full realization of the Supreme Personality of Godhead, for the all-pervading nature of the Supreme Lord is but a partial representation of His opulence.

Even the transcendental relationship experienced by a devotee of Narayana in Vaikuntha is incomplete in that it is not realization of a relationship with Krishna in Goloka Vrndavana. The devotees of Krishna do not relish devotional service to Narayana because devotional service to Krishna is so attractive that Krishna's devotees do not desire to worship any other form. Thus the gopis of Vrndavana do not like to see Krishna as the husband of Rukmini, nor do they address Him as Rukminiramana.

In Vrndavana Krishna is addressed as RadhaKrishna, or Krishna, the property of Radharani. Although the husband of Rukmini and Radha's Krishna are on the same level in the ordinary sense, still, in the spiritual world, the names indicate different understandings of various aspects of Krishna's transcendental personality. If one equalizes Rukminiramana, Radharamana, Narayana or any other name of the Supreme Lord, he commits the fault of overlapping tastes, which is technically called rasabhasa. Those who are expert devotees do not accept such amalgamations which are against the conclusions of pure devotional service.

Although Sri Krishna, the Supreme Personality of Godhead, embodies all superexcellence and beauty, when He is amongst the damsels of Vraja, He is known as Gopijanavallabha. The devotees cannot relish the beauty of the Supreme Lord more than the damsels of Vraja. In Srimad-Bhagavatam (10.33.7) it is confirmed that although Krishna, the son of Devaki, is the last word in superexcellence and beauty, when He is amongst the gopis it appears that He is a sublime jewel set amongst divine golden craftsmanship. Although Lord Chaitanya accepted this as the highest realization of the Supreme Lord as conjugal lover, He nonetheless requested Ramananda Raya to proceed further.

Upon hearing this request, Ramananda Raya remarked that this was the first time that he had been asked to go further than the gopis in an attempt to understand Krishna. There is certainly transcendental intimacy between the damsels of Vraja and Krishna, Ramananda pointed out, but out of all the relationships, the relationship between Radharani and Krishna in conjugal love is the most perfect. No common man can understand the ecstasy of transcendental love between Radharani and Krishna, nor can he understand the transcendental flavour of the transcendental love between Krishna and the gopis. Yet if one tries to follow in the footsteps of the gopis, he may become situated in the highest stage of transcendental love. Thus one who wants to be elevated to the transcendental stage of perfection should follow in the footsteps of the damsels of Vraja as an assistant maidservant of the gopis.

Lord Chaitanya exhibited the mode of Srimati Radharani when She was contacted from Dwarka by Sri Krishna. Such transcendental love is not possible for any common man; therefore one should not imitate the highest perfectional stage exhibited by Chaitanya Mahaprabhu. If, however, one desires to be in that association, he may follow in the footsteps of the gopis. In the Padma Purana it is stated that just as Radharani is dear to Krishna, similarly the kunda known as Radhakunda is also very dear to Him. Radharani is the only gopi who is dearer to Krishna than ail the other gopis. In Srimad-Bhagavatam (10.30.28) it is also stated that Radharani and the gopis render the highest perfectional loving service to the Lord and that the Lord is so pleased with them that He does not wish to leave the company of Srimati Radharani.

When Lord Chaitanya heard Ramananda Raya speak of the loving affairs between Krishna and Radharani, He said, "Please go further. Go on and on." The Lord also said that He was enjoying with great relish the descriptions of the loving affairs between Krishna and the gopis. "It is as if a river of nectar is flowing from your lips," He said. Ramananda Raya continued to point out that when Krishna danced amongst the gopis He thought, "I am not giving any special attention to Radharani." Because amongst the other gopis Radharani was not so much an object of special love, Krishna stole Her away from the area of the rasa dance and showed Her special favour. After explaining this to Lord Chaitanya, Ramananda Raya said, "Now let us relish the transcendental loving affairs between Krishna and Radha. These have no comparison in this material world."

Ramananda Raya thus continued his descriptions. During one performance of the rasa dance, Radharani suddenly left the area, as if She were angry that no special attention was being shown Her. Krishna was desirous of seeing Radharani in order to fulfil the purpose of the rasa dance, but not seeing Radharani there, He became very sorrowful and went to search Her out. In Gita-govinda there is a verse which states that the enemy of Kamsa, Krishna, also wanted to be entangled in love affairs with women and thus simply took Radharani away and left the company of the other damsels of Vraja. Krishna was very much afflicted by Radharani's absence and, being thus mentally distressed, began to search Her out along the banks of the Yamuna.

Failing to find Her, He entered the bushes of Vrndavana and began to lament. Ramananda Raya pointed out that when one discusses the purport of these two special verses of Gita-govinda (3.1-2), he can relish the highest nectar of Krishna's and Radha's loving affairs. Although there were many gopis to dance with Krishna, Krishna especially wanted to dance with Radharani. In the rasa dance Krishna expanded Himself and placed Himself between every two gopis, but He was especially present with Radharani. However, Radharani was not pleased with Krishna's behaviour.

As described in Ujjvala-nilamani: "The path of loving affairs is just like the movement of a snake. Amongst young lovers, there are two kinds of mentality—causeless and causal." Thus when Radharani left the area of the rasa dance out of anger at not receiving special treatment, Krishna became very sorrowful to see Her absent. The perfection of the rasa dance was considered to be complete due to Radharani's presence, and in Her absence Krishna considered the dance to be disrupted. Therefore He left the arena to search Her out. When He could not find Radharani after wandering in several places, He became very distressed. Thus it is understood that Krishna could not enjoy His pleasure potency even in the midst of all the gopis. But in the presence of Radharani He was satisfied.

When this transcendental love between Radharani and Krishna was described by Ramananda Raya, Lord Chaitanya admitted, "I came to you to understand the transcendental loving affairs between Krishna and Radha, and now I am very satisfied that you have described them so nicely. I can understand from your version that this is the highest loving state between Krishna and Radha." Yet Lord Chaitanya still requested Ramananda Raya to explain something more: "What are the transcendental features of Krishna and Radharani, and what are the transcendental features of the reciprocation of Their feelings, and what is the love between Them? If you kindly describe all this to Me, I will be very much obliged. But for you, no one can describe such things."

"I do not know anything," Ramananda Raya replied in all humility. "I am simply saying what You are causing me to say. I know that You are Krishna Himself, yet You are relishing hearing about Krishna from me. Therefore please excuse me for my faulty expression. I am just trying to express whatever You are causing me to express."

"I am a Mayavadi sanyasi," Lord Chaitanya protested. "I have no knowledge of the transcendental features of devotional service. By the greatness of Sarvabhauma Bhattacharya My mind has become clear, and I am now trying to understand the nature of devotional service to Lord Krishna. Bhattacharya recommended that I see you in order to understand Krishna. Indeed, he said that Ramananda Raya is the only person who knows something about love of Krishna. Therefore I have come to you upon the recommendation of Sarvabhauma Bhattacharya. Please, then, do not hesitate to relate to Me all the confidential affairs between Radha and Krishna."

In this way Lord Chaitanya actually took the subordinate position before Ramananda Raya. This has very great significance. If one is serious about understanding the transcendental nature of Krishna, he should approach a person who is actually enriched with Krishna consciousness. One should not be proud of his material birth, material opulence, material education and beauty and with these things try to conquer the mind of an advanced student of Krishna consciousness. One who thus goes to a Krishna conscious person, thinking that he would be favorably induced, is deluded about this science. One should approach a Krishna conscious person with all humility and put relevant questions to him. If one goes to challenge him, such a highly elevated Krishna conscious person would not be available for any service. A challenging puffed-up person cannot gain anything from a Krishna conscious man; he would simply remain in material consciousness. Although Lord Chaitanya was born in a high brahmana family and was situated in the highest perfectional stage of sannyasa, He nonetheless showed by His behaviour that even an elevated person would not hesitate to take lessons from Ramananda Raya, although Ramananda appeared as a householder situated in a social status beneath that of a brahmana.

Thus Lord Chaitanya clearly showed that a sincere student never cares whether his spiritual master is born in a high brahmana family or Kshatriya family, or whether he is a sanyasi, brahmacari or whatever. Whoever can teach one about the science of Krishna is to be accepted as guru.

The Supreme Perfection

Whatever position one may have, if he is fully conversant with the science of Krishna, Krishna consciousness, he can become a bona fide spiritual master, initiator or teacher of the science. In other words, one can become a bona fide spiritual master if he has sufficient knowledge of the science of Krishna, Krishna consciousness. The position does not depend on a particular position in society or on birth. This is the conclusion of Lord Chaitanya Mahaprabhu, and it is in accordance with the Vedic injunctions.

On the strength of this conclusion, Lord Chaitanya, previously known as Vishvambhara, accepted a spiritual master, Isvara Puri, who was a sanyasi. Similarly, Lord Nityananda Prabhu and Sri Advaita Acharya also accepted another sanyasi as their spiritual master, Madhavendra Puri. This Madhavendra Puri is also known as Laksmipati Tirtha. Similarly, another great acharya, Sri Rasikananda, accepted Sri Syamananda as his spiritual master, although he was not born of a brahmana family. So also Ganganarayana Cakravarti accepted Narottama dasa Thakura as spiritual master. In ancient days there was even a hunter named Dharma, who became a spiritual master for many people. There are clear instructions in Mahabharata and Srimad-Bhagavatam (7.11.32) stating that a person—be he brahmana, Kshatriya, vaisya or sudra—should be accepted by his personal qualifications and not by birth.

One's position should be established by personal qualifications and not by birth. For example, if a man is born in a brahmana family but his personal qualifications are those of a sudra, he should be accepted as a sudra. Similarly, if a person is born in a sudra family but has the qualifications of a brahmana, he should be accepted as a brahmana. All sastric injunctions, as well as the versions of great sages and authorities, establish that a bona fide spiritual master is not necessarily a brahmana. The only qualification is that he be conversant with the science of Krishna, Krishna consciousness. That alone makes one perfectly eligible to become a spiritual master. This is the conclusion of Sri Chaitanya Mahaprabhu in His discussions with Ramananda Raya.

In the Hari-bhakti-vilasa it is stated that if one bona fide spiritual master is born in a brahmana family and another, who is also qualified, is born in a sudra family, one should accept the one who is born in a brahmana family. This statement serves as a social compromise, but it has nothing to do with spiritual understanding. This injunction is only applicable for those who consider social status more important than spiritual status. It is not for people who are spiritually serious. A serious person would accept Chaitanya Mahaprabhu's instruction that anyone—regardless of his position—conversant with Krishna science must be accepted as the spiritual master.

There are many injunctions in the Padma Purana which state that a highly elevated spiritually advanced devotee of the Lord is always a first-class devotee and is therefore a spiritual master, but a highly elevated person born in a brahmana family cannot be a spiritual master unless he is a devotee of the Lord. A person born in a brahmana family may be conversant with all of the rituals of the Vedic scriptures but if he is not a pure devotee, he cannot be a spiritual master. In all shastras the chief qualification of a bona fide spiritual master is that he be conversant in the science of Krishna.

Lord Chaitanya therefore requested Ramananda Raya to go on teaching Him without hesitation, not considering Lord Chaitanya's position as a sanyasi. Thus Lord Chaitanya urged him to continue speaking on the pastimes of Radha and Krishna.

"Because You are asking me to speak of the pastimes of Radha and Krishna," Ramananda Raya humbly submitted, "I will obey Your order. I will speak in whatever way You like." Thus Ramananda Raya humbly submitted himself as a puppet before Lord Chaitanya, the puppet master. He only wanted to dance according to the will of Chaitanya Mahaprabhu. He compared his tongue to a stringed instrument, saying, "You are the player of that instrument." Thus as Lord Chaitanya would play, Ramananda Raya would vibrate the sound.

He said that Lord Krishna is the Supreme Personality of Godhead, the source of all incarnations and the cause of all causes. There are innumerable Vaikuntha planets, innumerable incarnations, expansions of the Supreme Lord, and innumerable universes also, and of all these existences the Supreme Lord Krishna is the only source. His transcendental body is composed of eternity, bliss and knowledge, and He is known as the son of Maharaja Nanda and the inhabitant of Goloka Vrndavana. He is full with six opulences—all wealth, strength, fame, beauty, knowledge and renunciation.

In Brahma-samhita (5.1) it is confirmed that Krishna is the Supreme Lord, the Lord of all lords, and His transcendental body is sac-cid-ananda. No one is the source of Krishna, but Krishna is the source of everyone. He is the supreme cause of all causes and the resident of Vrndavana. He is also very attractive, just like Cupid. One can worship Him by kama-Gayatri mantra.

In Brahma-samhita the transcendental land of Vrndavana is described as being always spiritual. That spiritual land is populated by goddesses of fortune, who are known as gopis. These are all beloved of Krishna, and Krishna is the only lover of all those gopis. The trees of that land are kalpa-vrksa, wish fulfilling trees, and one can have anything he wants from them. The land is made of touchstone and the water of nectar. In that land all speech is song, and all walking is dancing, and one's constant companion is the flute. Everything is self-luminous, just like the sun in this material world. The human form of life is meant for understanding this transcendental land of Vrndavana, and one who is fortunate should cultivate knowledge of Vrndavana and its residents. In that supreme abode are surabhi cows that overflood the land with milk. Since not even a moment there is misused, there is no past, present or future. An expansion of this Vrndavana, which is the supreme abode of Krishna, is also present on this earth, and superior devotees worship it as the supreme abode.

However, no one can appreciate Vrndavana without being highly elevated in spiritual knowledge, Krishna consciousness. According to ordinary experience, Vrndavana appears to be just like an ordinary village, but in the eyes of a highly elevated devotee, it is as good as the original Vrndavana. A great saintly acharya has sung: "When will my mind be cleared of all contamination so I will be able to see Vrndavana as it is? And when will I be able to understand the literatures left by the Gosvamis so that I will be able to know of the transcendental pastimes of Radha and Krishna?"

The loving affairs between Krishna and the gopis in Vrndavana are also transcendental. They appear as ordinary lusty affairs of this material world, but there is a gulf of difference. In the material world there may be the temporary awakening of lust, but it disappears after so-called satisfaction. In the spiritual world the love between the gopis and Krishna is constantly increasing. That is the difference between transcendental love and material lust. The lust, or so-called love, arising out of this body is as temporary as the body itself, but the love arising from the eternal soul in the spiritual world is on the spiritual platform, and that love is also eternal. Therefore Krishna is addressed as the ever green Cupid.

Lord Krishna is worshipped by the Gayatri mantra, and the specific mantra by which He is worshipped is called kama-Gayatri. Vedic literatures explain that, that sound vibration which can elevate one from mental concoction is called Gayatri. The kama-Gayatri mantra is composed of 24 1.2 syllables thus:

klim kama-devaya vidmahe
puspa-banaya dhimahi
tanno 'nangah pracodayat

This kama-Gayatri is received from the spiritual master when the disciple is advanced in chanting Hare Krishna, Hare Krishna, Krishna Krishna, Hare Hare. Hare Rama, Hare Rama, Rama Rama, Hare Hare. In other words, this kama-Gayatri mantra and samskara, or reformation of a perfect brahmana, are offered by the spiritual master when he sees that his disciple is advanced in spiritual knowledge. Even then, the kama-Gayatri is not uttered under certain circumstances. In any case, the chanting of Hare Krishna is sufficient to elevate one to the highest spiritual platform.

In Brahma-samhita a nice description of the flute of Krishna is given: "When Krishna began to play on His flute, the sound vibration entered into the ear of Brahma as the Vedic mantra om" This om is composed of three letters—A, U, and M—and it describes our relationship with the Supreme Lord, our activities by which we can achieve the highest perfection of love and the actual position of love on the spiritual platform. When the sound vibration of Krishna's flute is expressed through the mouth of Brahma, it becomes Gayatri. Thus by being influenced by the sound vibration of Krishna's flute, Brahma, the supreme creature and first living entity of this material world, was initiated as a brahmana. That Brahma was initiated as a brahmana by the flute of Krishna is confirmed by Srila Jiva Gosvami. When Brahma was enlightened by the Gayatri mantra through Krishna's flute, he attained all Vedic knowledge. Acknowledging the benediction offered to him by Krishna, he became the original spiritual master of all living entities.

The word klim added to the Gayatri mantra is explained in Brahma-samhita as the transcendental seed of love of Godhead, or the seed of the kama-Gayatri. The object is Krishna, who is the ever green Cupid, and by utterance of klim mantra Krishna is worshipped. It is also stated in the Gopala-tapani Upanishad that when Krishna is spoken of as Cupid, one should not think of Him as the Cupid of this material world. As already explained, Vrndavana is the spiritual abode of Krishna, and the word Cupid is also spiritual and transcendental. One should not take the material Cupid and Krishna to be on the same level. The material Cupid represents the attraction of the external flesh and body, but the spiritual Cupid is the attraction by which the Supersoul attracts the individual soul. Actually lust and sex are there in spiritual life, but when the spirit soul is embodied in material elements, that spiritual urge is expressed through the material body and is therefore pervertedly reflected. When one actually becomes conversant in the science of Krishna consciousness, he can understand that his material desire for sex is abominable, whereas spiritual sex is desirable.

Spiritual sex is of two kinds: one in accordance with the constitutional position of the self and the other in accordance with the object. When one understands the truth about this life but is not completely cleansed of material contamination, he is not factually situated in the transcendental abode, Vrndavana, although he may understand spiritual life. When, however, one becomes free from the sex urges of the material body, he can actually attain the supreme abode of Vrndavana. When one is so situated, he can utter the kama-Gayatri and kama-bija mantra.

Ramananda Raya then explained that Krishna is attractive both for men and women, for the movable and the immovable—indeed, for all living entities. For this reason He is called the transcendental Cupid. Ramananda Raya then quoted a verse from Srimad-Bhagavatam (10.32.2) stating that when the Lord appeared before the damsels of Vraja smiling and playing on His flute, He appeared just like Cupid.

There are different kinds of devotees who have different aptitudes and relationships with the Supreme Lord. Any one relationship with the Lord is as good as any other because the central point is Krishna. As stated in Bhakti-rasamrta-sindhu: "Krishna is the reservoir of all pleasures, and He is always attracting the gopis by the spiritual luster of His body. He especially attracts Taraka, Pali, Syama, and Lalita. Krishna is very dear to Radharani, the foremost gopi." Like Krishna, the gopis are glorified by Krishna's pastimes. There are different kinds of relationships with Krishna, and anyone who is attracted to Krishna by a particular relationship is glorified.

Krishna is so beautiful, transcendental and attractive that He sometimes attracts even Himself. The following verse appears in Gita-govinda (1.11):

> *visvesam anuranjanena janayann anandam indivara-*
> *sreni-syamala-komalair upanayann angair anangotsavam*
> *svacchandam vraja-sundaribhir abhitah pratyangam alingitah*
> *srngarah sakhi murtiman iva madhau mugdho harih kridati*

"My dear friend, just see how Krishna is enjoying His transcendental pastimes in the spring by expanding the beauty of His personal body. His soft legs and hands, just like the most beautiful moon, are used on the bodies of the gopis. When He embraces different parts of their bodies, He is so beautiful. Krishna is so beautiful that He attracts even Narayana, as well as the goddess of fortune who associates with Narayana."

In Srimad-Bhagavatam (10.89.58) the Bhuma-purusa (Maha-Vishnu) told Krishna, "My dear Krishna and Arjuna, I have taken the brahmana's sons just to see You." Arjuna attempted to save some youths who had died untimely at Dwarka, and when he failed to save them, Krishna took him to the Bhuma-purusa, and when the Bhuma-purusa brought forth those dead bodies as living entities, He said, "Both of you appear to preserve religious principles in the world and to annihilate the demons."

In other words, the Bhuma-purusa was also attracted by the beauty of Krishna, and He concocted this pastime just as a pretext to see Krishna. It is recorded in Srimad-Bhagavatam (10.16.36) that after the serpent Kaliya was punished by Krishna, one of Kaliya's wives told Krishna, "Dear Lord, we cannot understand how this fallen serpent got the opportunity of being kicked by Your lotus feet when even the goddess of fortune underwent austerities for several years just to see You."

How Krishna is attracted by His own beauty is described in Lalita-madhava (8.34). Upon seeing His own picture, Krishna lamented, "How glorious this picture is! It is attracting Me just as it attracts Radharani."

After giving a summary of the beauty of Krishna, Ramananda Raya began to speak of the spiritual energy of Krishna, which is headed by Srimati Radharani. Krishna has immense energetic expansions. Three energies are predominant: the internal energy, external energy and marginal energy. This is confirmed in the Sixth Chapter of Vishnu Purana where it is said that Vishnu has one energy, which is called spiritual energy, and it is manifested in three ways. When spiritual energy is overwhelmed by ignorance, it is called marginal energy. As far as spiritual energy itself is concerned, it is exhibited in three forms because Krishna is a combination of eternity, bliss and knowledge. As far as His bliss is concerned, His spiritual energy is manifested as the pleasure giving potency. His eternity is manifested as energy, and His knowledge is manifested as spiritual perfection. As confirmed in Vishnu Purana (1.12.69): "The pleasure potency of Krishna gives Krishna transcendental pleasure and bliss." Thus when Krishna wants to enjoy pleasure, He exhibits His own spiritual potency known as hladini.

In His spiritual form, Krishna enjoys His spiritual energy, and that is the sum and substance of the Radha-Krishna pastimes. These pastimes can only be understood by elevated devotees. One should not try to understand the Radha-Krishna potencies and pastimes from the mundane platform. Generally people misunderstand these as being material.

When the pleasure potency is further condensed, it is called mahabhava. Srimati Radharani, the eternal consort of Krishna, is the personification of that mahabhava. In this regard, in Ujjvala-nilamani (4.3) Rupa Gosvami states that there are two competitors in love with Krishna, Radharani and Candravali. When they are compared, it appears that Radharani is superior, for She possesses mahabhava-svarupa. Mahabhava-svarupa, the personification of mahabhava, is applicable to Radharani only. Mahabhava is full of the pleasure potency, and it is an exhibition of the highest love for Krishna. Radharani is therefore known throughout the world as the most beloved of Krishna, and Her name is always associated with Krishna as Radha-Krishna.

Brahma-samhita (5.37) also confirms that Krishna expands Himself by His pleasure potency in the spiritual world and that these potencies are all nondifferent from Krishna. Although Krishna is always enjoying the company of His pleasure potency expansions, He is all-pervading. Thus Brahma offers his respectful obeisances to Govinda, the cause of all causes.

As Krishna is the highest emblem of spiritual perfection, so Radharani is the highest emblem of that spiritual pleasure potency by which Krishna is satisfied. Since Krishna is unlimited, in order to satisfy Him Radharani is also unlimited. Krishna is satisfied just by seeing Radharani, but Radharani expands Herself in such a way that Krishna desires to enjoy

Her more. Because Krishna was unable to estimate the pleasure potency of Radharani, He decided to accept the role of Radharani, and that combination is Sri Chaitanya Mahaprabhu.

Ramananda Raya then began to explain Radharani as the supreme emblem of Krishna's pleasure potency. Radharani expands Herself in different forms, known as Lalita, Vishakha and Her other confidential associates. In his book Ujjvala-nilamani, Rupa Gosvami explains the characteristics of Srimati Radharani. He points out that the body of Radharani is in itself an actual evolution of transcendental pleasure. That body is decorated with flowers and fragrant aromas and is complete with transcendental love for Krishna.

That is the personification of His pleasure potency. That transcendental body takes bath three times: first in the water of mercy, second in the water of youthful beauty, and third in the water of youthful luster. After taking bath three times in that way, Her body is covered with shining garments and decorated with Krishna's personal beauty, which is compared to cosmetics. Thus Her beauty constitutes the highest artistry. Her body is also decorated with the ornaments of spiritual ecstasy—trembling tears, petrification, perspiration, choking, cessation of all bodily functions due to transcendental pleasure, stumbling, high blood pressure and madness.

The decorative transcendental pleasure potency manifests nine symptoms. Five of these are manifested by the expansion of Her personal beauty, which is adorned with garlands of flowers. Her patient calmness is compared with a covering of cloths which have been cleansed by camphor. Her confidential agony for Krishna is the knot in Her hair, and the mark of tilaka on Her forehead is Her fortune. Radharani's sense of hearing is eternally fixed on Krishna's name and fame. One's lips become red from chewing betel nuts.

Similarly, the borders of Radharani's eyes are blackened due to Her complete attachment to Krishna. This darkness might be compared to ointment used by nature when nature jokes with Radha and Krishna. Radharani's smile is just like the taste of camphor. The garland of separation moves on Her body when She lies down on the bed of pride within the room of aroma. Out of ecstatic affection for Krishna, Her breast is covered by the blouse of anger. Reputed as the best of all Krishna's girl friends, She plays a stringed instrument. When Krishna stands in His youthful posture, She puts Her hand on His shoulder. Although She possesses so many transcendental qualities, She is always engaged in the service of Krishna.

Srimati Radharani is decorated with suddipta-sattvika emotions, which sometimes include tribulation and sometimes pacification. All the transcendental ecstasies are manifested in the body of Srimati Radharani. Suddipta-sattvika emotions are manifest when a lover is overwhelmed with certain feelings which he or she cannot check. Radharani has another emotion called kilakincita, which is manifest in twenty different ways. These emotions are manifested partly due to body, mind and habit. As far as the bodily emotions are concerned, they are manifested in posture and movement. As far as the emotions of the mind are concerned, they are manifested as beauty, luster, complexion, flavour, talking, magnanimity and patience. As far as habitual emotions are concerned, they are manifested as pastimes, enjoyment, preparation and forgetfulness.

The tilaka of fortune is on the forehead of Srimati Radharani, and She also has a locket of prema-vaicittya. prema-vaicittya is manifest when a lover and beloved meet and fear separation.

Srimati Radharani is fifteen days younger than Krishna. She always keeps Her hand on the shoulders of Her friends, and She always talks and thinks of pastimes with Krishna. She always offers Krishna a kind of intoxication by Her sweet talks, and She is always prepared to fulfil His desires. In other words, She supplies all the demands of Sri Krishna, and She possesses unique and uncommon qualities for Krishna's satisfaction.

In Govinda-lilamrta there is a verse which states: "Who is the breeding ground for Krishna's affection? The answer is that it is only Srimati Radhika. Who is Krishna's most lovable object? The answer is that it is only Srimati Radhika and no one else." Sheen in the hair, moisture in the eyes and firmness in the breasts are all qualities present in Srimati Radhika. Only Srimati Radhika is able to fulfil all the desires of Krishna. No one else can do so.

Satyabhama is another competitor of Srimati Radharani, but she always desires to come to the standard of Srimati Radharani. Radharani is so expert in all affairs that all the damsels of Vraja come to learn arts from Her. She is so extraordinarily beautiful that even the goddess of fortune and Parvati, the wife of Lord Shiva, desire elevation to Her standard of beauty. Arundhati, who is known as the most chaste lady in the universe, desires to learn the standard of chastity from Srimati Radharani. Since even Lord Krishna cannot estimate Radharani's highly transcendental qualities, it is not possible for an ordinary man to estimate them.

After hearing Ramananda Raya speak of the qualities of Radha Krishna, Lord Chaitanya desired to hear from him about the reciprocation of love between Them. Ramananda Raya described Krishna as dhira-lalita, a word which describes a person who is very cunning and youthful, who is always expert in joking, who is without anxiety and who is always subservient to his girl friend. Krishna is always engaged in love affairs with Radharani, and He takes to the bushes of Vrndavana to enjoy His lusty activities with Her. Thus He successfully carries out His lusty instincts. In Bhakti-rasamrta-sindhu it is stated:. "By His impudent and daring talks about sex indulgence, Krishna obliged Srimati Radharani to close Her eyes, and taking advantage of this, Krishna painted many pictures on Her breasts. These pictures served as subject matter for Radharani's friends to joke about. Thus Krishna was always engaged in lusty activities, and thus He made His youthful life successful."

Upon hearing of these transcendental activities, Lord Chaitanya said, "My dear Ramananda, what you have explained regarding the transcendental pastimes of Sri Radha and Krishna is perfectly correct, yet there is something more I would like to hear from you."

"It is very difficult for me to express anything beyond this," Ramananda Raya replied. "I can only say that there is an emotional activity called prema-vilasa-vivarta, which I may try to explain but I do not know whether You will be happy to hear it." In prema-vilasa there are two kinds of emotional activities—

separation and meeting. That transcendental separation is so acute that it is actually more ecstatic than meeting. Ramananda Raya was expert in understanding these highly elevated dealings between Radha and Krishna, and he composed a nice song which he narrated to the Lord. The purport of the song is that the lover and the beloved, before meeting, generate a kind of emotion by the exchange of their transcendental activities. That emotion is called raga, or attraction. Srimati Radharani expressed Her willingness that "this attraction and affection between Us rise to the highest extent," but the cause of this attraction is Radharani Herself. "Whatever the cause may be," Radharani said, "that affection between You and Me has mixed Us in oneness. Now that it is the time of separation, I cannot see the history of the evolution of this love. There was no cause or mediator in Our love save Our meeting itself and the visionary exchange of feelings."

This exchange of feelings between Krishna and Radharani is very difficult to understand unless one is elevated to the platform of pure goodness. Such transcendental reciprocation is not even possible to understand from the platform of material goodness. One has to actually transcend material goodness in order to understand. This is because the exchange of feelings between Radha and Krishna is not a subject matter of this material world. Even the greatest mental speculator cannot understand this, directly or indirectly. Material activities are manifested either for the gross body or the subtle mind, but this exchange of feelings between Radha and Krishna is beyond such manifestations and beyond intellectual mental speculation. It can only be understood by purified senses freed from all the designations of the material world.

Those who have purified senses can understand these transcendental features and exchanges, but those who are impersonalists and who have no knowledge of spiritual senses can only discriminate within the scope of the material senses and thus cannot understand spiritual exchanges or spiritual-sensual activities. Those who are elevated by virtue of experimental knowledge can only satisfy their blunt material senses either by gross bodily activities or by mental speculation.

Everything generated from the body or the mind is always imperfect and perishable, but transcendental spiritual activities are always bright and wonderful. Pure love on the transcendental platform is the paragon of purity devoid of material affection and completely spiritual. Affection for matter is perishable, as indicated by the inebriety of sex in the material world, but there is no such inebriety in the spiritual world. Hindrances on the path of sense satisfaction cause material distress, but one cannot compare that with spiritual separation. In spiritual separation there is neither inebriety nor ineffectiveness, as one finds with material separation.

Lord Chaitanya admitted that this is the highest position of transcendental loving reciprocation, and He told Ramananda Raya, "By your grace only have I been able to understand such a high transcendental position. Such a position cannot be attained without the performance of transcendental activities. So will you kindly explain to Me how I can raise Myself to this platform?"

"It is similarly difficult for me to make You understand," Ramananda replied. "As far as I am concerned, I can only speak what You wish me to. No one can escape Your supreme will. Indeed, there is no one in the world who can surpass Your supreme will, and although I appear to be speaking, I am actually not the speaker. You are speaking. Therefore You are both the speaker and the audience. Thus let me speak only as You will me to speak about the performance required to attain this highest transcendental position."

Ramananda Raya then began to relate the confidential and transcendental activities of Radha and Krishna. These activities cannot be understood in the emotional relationship with the Supreme Lord as master and servant, friend and friend or parent and son. This confidential subject matter can be understood only in the association of the damsels of Vraja, for the confidential activities have arisen from the feelings and emotions of those damsels. Without the association of the damsels of Vraja, one cannot nourish or cherish such transcendental understanding. In other words, these confidential pastimes of Radha and Krishna have expanded through the mercy of the damsels. Without their mercy, they cannot be understood. One has to follow in the footsteps of the damsels of Vraja in order to understand.

When one is actually situated in that understanding, he becomes eligible to enter into the confidential pastimes of Radha and Krishna. There is no alternative to understand Their confidential pastimes. This is confirmed in Govinda-lilamrta (10.17): "Although manifest, happy, expanded and unlimited, the emotional exchanges between Radha and Krishna can only be understood by the damsels of Vraja or by their followers." Just as no one can understand the expansion of the spiritual energy of the Supreme Lord without His causeless mercy, no one can understand the transcendental sex life between Radha and Krishna without following in the footsteps of the damsels of Vraja.

The personal associates of Radharani are called sakhis, and Her near assistants are called manjaris. It is very difficult to express their dealings with Krishna because they have no desire to mix with Krishna or to enjoy Him personally. Rather, they are always ready to help Radharani associate with Krishna. Their affection for Krishna and Radharani is so pure that they are simply satisfied when Radha and Krishna are together. Indeed, their transcendental pleasure is in seeing Radha and Krishna united. The actual form of Radharani is just like a creeper embracing the tree of Krishna, and the damsels of Vraja, the associates of Radharani, are just like the leaves and flowers of that creeper. When a creeper embraces a tree, the leaves and flowers as well as the creeper automatically embrace it. Govinda-lilamrta (10.16) confirms that Radharani is the expansion of the pleasure potency of Krishna and is compared with a creeper, and Her associates, the damsels of Vraja, are compared to the flowers and leaves of that creeper. When Radharani and Krishna enjoy Themselves, the damsels of Vraja relish the pleasure more than Radharani Herself.

Although the associates of Radharani do not expect any personal attention from Krishna, Radharani is so pleased with them that She arranges individual meetings between Krishna and the damsels of Vraja. Indeed, Radharani tries to combine or unite Her associates with Krishna by many transcendental maneuvers, and She takes more pleasure in these meetings than in Her own meetings with Him. When Krishna sees that both Radharani and Her associates are pleased by His association, He becomes more satisfied.

Such association and loving reciprocation have nothing to do with material lust, although it resembles the material union between man and woman. It is only because that similarity is there that such reciprocation is sometimes called, in transcendental language, transcendental lust. As explained in Gautamiya-tantra (Bhakti-rasamrta-sindhu 1.2.285): "Lust means attachment to one's personal sense gratification. But as far as Radharani and Her associates are concerned, they did not desire personal sense gratification. They only wanted to satisfy Krishna." This is further confirmed in the Srimad-Bhagavatam (10.31.19) in one of the speeches of the gopis:

yat te sujata-caranamburuham stanesu
bhitah sanaih priya dadhimahi karkasesu
tenatavim atasi tad vyathate na kim svit
kurpadibhir bhramati dhir bhavad-ayusam nah

"My dear friend Krishna, You are now roaming in the forest with Your bare feet, which You sometimes keep on our breasts. When Your feet are on our breasts, we think that our breasts are too hard for Your soft feet. Now You are wandering in the forest and walking over stones, and we do not know how You are feeling. Since You are our life and soul, the displeasure You undergo in travelling over rough stones is giving us great distress." Such feelings expressed by the damsels of Vraja constitute the highest Krishna conscious emotions. Anyone who actually becomes captivated by Krishna consciousness approaches this level of the gopis. There are sixty-four categories of devotional service by performance of which one can rise to the gopis' stage of unconditional devotion. Affection for Krishna on the level of the gopis is called raganuga, spontaneous love. When one enters into a spontaneous loving affair with Krishna, there is no need to follow the Vedic rules and regulations.

There are various kinds of personal devotees of Lord Krishna in the transcendental abode. For example, there are servants of Krishna like Raktaka and Patraka and friends of Krishna just like Sridama and Sudama. There are also parents of Krishna, just like Nanda and Yashoda, who are also engaged in the service of Krishna according to their respective transcendental emotions. One who desires to enter into the supreme abode of Krishna can take shelter of one of such transcendental servitors. Then, through the execution of loving service, one can attain transcendental affection for Krishna. In other words, the devotee in this material world who executes loving service in pursuance of the activities of those eternal associates with Krishna also attains the same post when he is perfected.

The sages mentioned in the Upanishads and in sruti also desire the post of the gopis, and they also follow in the footsteps of the gopis in order to attain that highest goal of life. This is confirmed in Srimad-Bhagavatam (10.87.23) when it is said that the sages practice pranayama (trance) by controlling the breathing process, mind and senses through the mystic yoga practice. Thus they try to merge into the Supreme Brahman. This same goal is attained by atheists who deny the existence of God. If such atheists are killed by an incarnation of the Supreme Personality of Godhead, they also merge into the Brahman existence of the Supreme Lord. However, when the damsels of Vrndavana worship Sri Krishna, it is as though they have been bitten by a snake, for Krishna's body is compared with the body of a snake. A snake's body is never straight; it is always curling.

Similarly, Krishna often stands in a three curved posture, and He has bitten the gopis with transcendental love. The gopis are certainly better situated than all mystic yogis and others who desire to merge into the Supreme Brahman. The sages of Dandakaranya also follow in the footsteps of the damsels of Vraja in order to attain a similar position. One cannot attain that position simply by following the regulative principles. Rather, one must seriously follow the principles of the gopis. This is confirmed in Srimad-Bhagavatam (10.9.21) wherein it is stated that Lord Sri Krishna, the son of Srimati Yashoda, is not easily available to those following the principles of mental speculation but is easily available to all kinds of living beings who follow the path of devotional service.

There are many pseudo-devotees, claiming to belong to Lord Chaitanya Mahaprabhu's sect, who artificially dress themselves as the damsels of Vraja, and this is not approved by advanced spiritualists or advanced students of devotional service. Such people dress the outward material body because they foolishly confuse the body with the soul. They are mistaken when they think that the spiritual bodies of Krishna, Radharani and Their associates, the damsels of Vraja, are composed of material nature. One should know perfectly well that all such manifestations are expansions of eternal bliss and knowledge in the transcendental world.

They have nothing to do with these material bodies; thus the bodies, dresses, decorations and activities of the damsels of Vrndavana are not of this material cosmic manifestation. The damsels of Vrndavana are not a subject for the attraction of those in the material world; they are transcendental attractions for the all-attractive Krishna. Because the Lord is all-attractive, He is called Krishna, but the damsels of Vrndavana are attractive even to Krishna. Therefore they are not of this material world.

If one wrongly thinks that the material body is as perfect as the spiritual body and begins to imitate the damsels of Vrndavana, he becomes infested with the Mayavadi impersonal philosophy. The impersonalists recommend a process of aham grahopasana by which one worships his own body as the Supreme. Thinking in this way, such pseudo-transcendentalists dress themselves as the damsels of Vraja. Such activities are not acceptable in devotional service. Even Srila Jiva Gosvami, the most authoritative acharya in the Gaudiya sampradaya, has condemned these imitators. The process of transcendental realization is to follow in the footsteps of the associates of the Supreme Lord; therefore if one thinks himself to be a direct associate of the Supreme Lord, he is condemned. According to authorized Vaishnava principles, one should follow a particular devotee, and not think of himself as Krishna's associate.

In this way Ramananda Raya explained that one should accept the mood of the damsels of Vraja. In Chaitanya-caritamrta it is clearly said that one should accept the emotional activities and not imitate the dress of the associates of Krishna. One should also always meditate upon the affairs between Radha and Krishna in the transcendental world. One should think of Radha and Krishna twenty-four hours a day and eternally engage in Their service. One need not externally change his dress. By following the mood of the associates and friends of Radharani, one can ultimately achieve the perfectional stage and be transferred to Goloka Vrndavana, the transcendental abode of Krishna.

The mood of the gopis' emotional pursuit is called siddha-deha. This word indicates the pure spiritual body which is beyond the senses, mind and intelligence. Siddha-deha is the purified soul who is just suitable to serve the Supreme Lord. No one can serve the Supreme Lord as His associate without being situated in his pure spiritual identity. That identity is completely free from all material contamination. As stated in Bhagwad Gita, a materially contaminated person transmigrates to another material body by material consciousness. At the time of death he thinks materially and is therefore transferred into another material body. Similarly, when one is situated in his pure spiritual identity and thinks of the spiritual loving service rendered to the Supreme Lord, he is transferred to the spiritual kingdom to enter into the association of Krishna. In other words, by attaining one's spiritual identity and thinking of Krishna's associates, one can become qualified to transfer to the spiritual kingdom. No one can contemplate or think of the activities of the spiritual kingdom without being situated in his pure, spiritual identity (siddha). Thus Ramananda Raya said that without attaining siddha-deha, one can neither become an associate of the damsels nor render service directly to the Personality of Godhead, Krishna, and His eternal consort Radharani. In this regard, Ramananda quoted Srimad-Bhagavatam (10.47.60):

nayam sriyo 'nga u nitantarateh prasadah
svar-yositam nalina-gandha-rucam kuto 'nyah

rasotsave 'sya bhujadanda-grhita-kantha-
labdhasisam ya udagad vraja-vallavinam

"Neither the goddess of fortune, Lakshmi, nor even the damsels of the heavenly kingdom can attain the facilities of the damsels of Vrajabhumi—and what to speak of others."

Lord Chaitanya was very satisfied to hear these statements from Ramananda Raya, and He embraced him. Then both of them began to cry in the ecstasy of transcendental realization. Thus the Lord and Ramananda Raya discussed the transcendental pastimes of Radha and Krishna throughout the night, and in the morning they separated. Ramananda left to go to his place, and the Lord went to take His bath.

At the time of separation, Ramananda fell at the feet of Lord Chaitanya and prayed: "My dear Lord, You have come just to deliver me from this mire of nescience. Therefore I request that You remain here for at least ten days to purify my mind of this material contamination. There is no one else who can deliver such transcendental love of God."

"I have come to you to purify Myself by hearing from you the transcendental pastimes of Radha and Krishna," the Lord replied. "I am so fortunate, for you are the only teacher of such transcendental pastimes. I can find no one else in the world who can realize the transcendental loving reciprocation between Radha and Krishna. You are asking Me to stay here for ten days, but I feel like remaining with you for the rest of My life. Please come to Jagannath Puri, My headquarters, and we will remain together for the rest of our lives. Thus I can pass My remaining days in understanding Krishna and Radha by your association."

Sriman Ramananda Raya came again to see the Lord the next evening, and there were further discourses on this transcendental subject matter.

"What is the highest standard of education?" Lord Chaitanya began His inquiry, and Ramananda Raya immediately replied that the highest standard of education is knowledge of the science of Krishna. The standard of material education is sense gratification, but the highest standard of spiritual education is knowledge of the science of Krishna. In Srimad-Bhagavatam (4.29.50) it is stated that that work which pleases the Supreme Personality of Godhead is the highest work, and that science or knowledge which places one in full Krishna consciousness is the highest knowledge. Similarly, Prahalad Maharaja, while instructing his childhood friends at school, also stated that hearing of the Lord, chanting, remembering, worshipping, praying, serving, making friends with Krishna, and offering everything to Him constituted the highest spiritual knowledge.

"And what is the greatest reputation one can have?" Lord Chaitanya asked Ramananda Raya, and Ramananda immediately replied that a person reputed to be Krishna conscious should be considered the most famous man in the world.

knowledge of the service of Krishna. In Śrīmad-Bhāgavatam (4.29.49) it is stated that that work which pleases the Supreme Personality of Godhead is the highest work, and that science or knowledge which places one in full Krishna consciousness is the highest knowledge. Similarly, Prahlāda Mahārāja, while instructing his schoolmates, also stated that learning of the Lord, chanting His name, worshipping, praying, serving, becoming friends with Him, and offering everything to Him constitute the highest type of knowledge.

"And what is the greatest fame among all fames?" Lord Chaitanya asked. Rāmānanda immediately replied that a person reputed to be Krishna's devotee should be considered the most famous man in the world.

4

Agamas and South Indian Vaishnavism

Every society has codes for ethics helpful to its well being. The standard of good conduct and moral behaviour enjoined by them vary from individual to individual and from one stratum of society to another. All such codes and standards among the Hindus derive from the Vedas.

Moral goodness is a question of behaviour. 1410. Everlasting happiness is the goal of good conduct. The law books (Smritis) and the conduct of those steeped in Vedic tradition are also authoritative 1411. Moral behaviour is alone held to contribute to individual progress and social welfare. Mere knowledge (even if it be philosophical or theological) can lead to no certain goal. 1412. The ethical codes are supported and stabilised by religious traditions. Hence the religious colouring given by Hindus even to secular activities.

The epics and the pecanas have played a significant role in fashioning moral concepts-particularly the Mahabharata, the Vishnudharmottara, the Vishnupurana, the Bhagwad Gita, and the Bhagavata. With the exception of the Gita, these texts reveal acquaintance with the vyiiha and sadguna aspects of God, which are vital to the Pancarata. The influence of the Sattvata-dharma (which is developed in the Mahabharata) may be traced to the books mentioned above. Still, the wide and varied aspects of Vaishnavism are so rich and complex as to deny any significant Agama influence on them. Dependence on God in leading a virtuous life is stressed in all ancient works.

Thus no one had any real freedom. God's (Vishnu's) will creates and sustains the world and provides the people with security, 1413. Uttering his name even light-heartedly makes for happiness here and hereafter; no need to speak of devotion or dedication of all actions to God bringing this about, 1414. God is the friend after all, high or low, 1415. His devotees never suffer 1416.

Devotion is of various kinds to physical and mental capacities, 1417. Its quality is determined by the spirit, and not by pomp or show. The greatest of all activities is worship of God, and adoration of His devotees is still more worthy, for God is more pleased when his devotees receive worship than when he does, 1418. Prapatti is declared to be the surest means of attaining God, 1419. The Kalpasutras and Smritis that deal with the social and religious aspects of life, explain how health and purity can be ensured. Bodhayana 1420 and others prescribe general rules for conduct and also give specific direction for the conduct of Vaishnavas and the worship of God.

The Smritis of Manu, 1421 Sandilya, 1422 Vasistha, 1424 and others are similar. The Agmas' impact on Vaishnavism seems to have become significant only from the time of Ramanuja. The earlier attitudes may perhaps be due to the Brahmasutra discussion on the validity of the Pancharatra and the stand of Shankara and Bhaskara that is this Agama is unacceptable in part. Yamuna's rejoinder eased the position. After Ramanuja incorporated their doctrines in his Gadyatraya, they came to occupy a position of unquestioned authority. But the Agamas have not influenced Vaishnava tradition to such an extent as to make it conform extensively to their ideals and practises. Ramanuja also wrote the Nitya on the way of worshipping God everyday at home.

It is based on the Pancharatra. Though the Vaikhanasa also has preseribed a daily routine, it is intended only for the followers of the Vaikhanasa-sutras, 1425. Hence, following the lines laid down by Ramanuja, his immediate disciples, (Srivatsankamistra and Sriranganarayana) brought forth their own manuals. They were followed by Parasarabhatta, Nanjiyar, Narayana-muni, Vedanta-desika, Varavara-muni and others.

Works on ahnika were written by several writers in later periods.

The fivefold division of the daily routine (pancha-kala-prakriya) forms the basis of these works. Till recently there were so many Srivaisnavas who scrupulously carried out this routine. That is not the case now. The indispensable qualification for being a Vaishnava is wearing the urdhva-pundra. The marks which Hindu men and women wear on their foreheads are known as pundras. They may be vertical (urdhva), horizontal (tiryak) or circular (vartula).

The Vaishnava tradition enjoins the vertical variety, on the authority of the Vedas. The urdhva-pundra is claimed to keep away evil spirits, 1426. It must be worn to ensure purity, while offering sandhya prayers, doing homa, worshipping God, studying the Vedas and such other religious activities, 1427. Otherwise, they will be futile, 1428. The material used for the pundra is white mud; 1429 sandal and some other things, recommended in different contexts, are held apply to only particular persons 1430. The white mud should be taken from only select spots like Sriangam, Venkatadri, Srimusnanm, Tirunarayanapuram, Vrisabhadri (Tirumalirunsolai), banks of rivers, ant hills, the areas where tulsi grows and holy places, 1431.

The urdhva-pundra has to start at the tip of the nose and rise upwards in parallel columns with some space in between them, where a red or yellow vertical line is drawn with the help of a powder made of turmeric and other ingredients 1432 and called "Sricurna". Several opinions are recorded about the shape of the urdhva-pundra. It may be of the form of flame, the leaf of the bamboo, a flower bud and Vishnu's foot or feet and so on and each form is said to achieve a specific purpose. But the practice among Srivasnavas is to have it in the form of the divine foot or feet. However, the two sects, Vadakalai and Tenkalai, have chosen to adopt slightly different forms of pundra. As a rule, red-coloured Sricurna is worn by both, but some of the former use the yellow colour. Vedantadesika discusses an interesting question about the form of pundra. On the idols of God in temples, the pundra is in the form of a flame. He says that the same form should not be used by men, for whom the form is that of Hari-pada. If they try to imitate the Lord in this, they will seek to follow His example in other matters as well?

The number must be twelve including on in the forehead. The others are to be put on specific parts of the body. There appears to be a somewhat weaker sanction for wearing only four. The Pancharatra enjoins the uttering of the twelve names of Vishnu, when putting them on, while the Vaikhanasa prescribes specific mantras instead. Putting the urdhva pundra on the idols in the temples is authorised, as well as on wells, mantapas, horses, elephants and vehicles, on the vessels used for worship, flags and other objects belonging to the temple. At the entrance to houses, the pundra used to be carved on doors or painted on plank figures–of the discuss on the right, the conch on the left, Garuda to the right of the discuss and Hanuman to the left of the conch. In modern days, wearing of the urdhva-pundra has practically disappeared among the younger generations.

The practice of wearing the urdhva-pundra seems to be very ancient. Vedanta-desika in his Saccaritra-raska gives profuse citations from Vedic texts like Katha-sakha and Baudhayana Kalpasiitras like the Matsya, Brahamanda, Markandeya, Varistha and Skanda, and Smritis like Vishnu-smriti and Smrtichandrika. The two Agamas also advocate it. Vedanta-desika quotes from the Paramesvara, Paramesthi, and Sanatkumara Samhitas, though many passages cited from the Crahma-ratra section of the last mentioned text are from three chapters which are now lost. A fivefold inditiation known as the pancha-samskara is enjoined on all the Srivaisnavas without distinction of caste or sex. This sacriment is intended to fit one to discharge one's duties to God and man. It consists of tapa, pundra naman, mantra and ijya. It is also known as samasrayana or resorting to (the preceptor as refuge).

Tapa (heating) refers to the marking of the left and right shoulder blades of the initiate with small heated pieces of metal of the form of the conch and the discus respectively. Vedanta-desika quotes texts from the Vedas, the Pancharatra and the puranas as authority for tapa. Two Vedic passages cited by him are given below...

> *Those who have realized Brahman declare that the Brahmin must bear the discus on the right arm and the conch on the left.*
>
> *O Vishnu, the learned have on the upper part of their arms the sanctifying conch and discus in order to cross over the ocean of worldly existence.*

Another Vedic text cited by Vedanta-desika declares: "He who bears on his body the mark of the heated discus of the omnipresent Vishnu which gives security to the army of the gods, goes to the place which is devoid of misery, shaking off the sins–the place which the recluses who are without attachment attain." The word translated, as 'on his body' is 'vapusa', the instrumental form of 'vapus' meaning 'body', There is nothing against taking it to mean "the upper part of the right arm" which will all be in consonance with other texts and hoary traditions. Those who do not have their bodies marked by the heated discuss cannot exhaust the karmas (*i.e.*, effects of their past actions): which condition is a necessary preliminary to salvation, 1441.

Another Vedic passage testifies to ancient sages bearing on their bodies the marks of the divine weapons: The marks (lingas) mentioned here refer to those of discus, conch, mace, sword and bow, the five weapons of Vishnu. They are believed to have been worn on the right and left arms, the forehead, the head and the heart respectively. Now only the first two are worn. Instead of marking the shoulders by tapa with the marks of discus and conch, mud and sandal paste are used for the purpose by Vaishnavas belonging to schools other than that of Ramanuja. The scriptural text from the Rigaveda-khila mentioned above states: bibiharti vapuse abhitaptam. This clearly envisages tapa.

Obviously, this sacrament is intended to purify the human body and render it fit for service to God. Fire is regarded as a more effective purifier than water,–particularly, 'fire' in the heated symbols of Vishṇu or His weapons. Such marks on the body render it fit for any religious act that the Vaishnava has to do for pleasing God. The acharya officiating in this sacrament has been accused of cruelty. He hurts the body of the disciple only for a noble purpose. He must be likened to the surgeons who use sharp instruments to operate on the patients. Parents again punish their children severely to induce them to be good, studious and industrious: this is done out of love.

Again, in certain regions branding is a remedy for evil or means of winning good luck. Certain diseases are treated in Saurastra by branding with a heated iron rod below the affected part of body. In Mithila, the mother in law and others apply the flame of a lamp to the knee of the newly wedded bride to ensure good fortune for her. The law-books of Vriddha Harita, Sandilya and Vasistha and the Vishnudharmottara enjoin this practice. Sandilya says that one's utensils and cattle are to be given the marks of the conch and the discus.

Those who oppose the sacrament of tapa rely on a passage in S'rauta-sutra which frowns on scars from wounds in the body caused in three ways. The three ways, are taken to be through the bow-string, engaging in battles or involvement in scuffles in gambling. Such activities must be regarded as forbidden or allowed only to the extent at which no scars are received or given. But they are claimed to refer to wearing the emblems of Brahma, Vishnu or Rudra so as to cause scars. Such interpretation is untenable: it goes against the authority of the Vedas, epics, Agamas and the bulk of the Smritis. The few Smriti texts quoted against the sacrament cannot avail against this formidable corpus of authority. It should be remembered that there are references also, as already mentioned, to be practice of wearing the emblems of all five weapons of Vishnu (the conch, discus, mace, sword and bow) to get freedom from samsara.

There is no decisive evidence to show how old the rituals of tapa as also the pancha-samskara generally is. It might have been originally intended as a protection against evil spirits. The new-born child used to be given a garland containing the emblems. Earrings of the form of the conch and discus used to be worn by devotees of Vishnu. In course of time, probably by 700 A.D, it became essential for a Vaishnava to have the samskara. In Nathamuni's days, it must have been in vogue, though not every Vaishnava of the times observed it. (Wearing urdhva-pundra might have become widespread earlier.) After Nathamuni, tradition records some instances of tapa being administrated. Tirumalai Nambi adminsitrated it to his two son in laws, and Periya Nambi to Ramanuja. After Ramanuja the sacrament of pancha-samskara became obligatory for all Srivaisnavas, irrepective of caste or sex. In a passage from the Mahabharata quoted earlier in this book, it is declared that Bhramins, Kshatriyas, Vaishyas and Sudras can engage in (ritualistic) worship of God if they are always devoted to their duties and are krta-laksanas, persons who have obtained marks of identity, that is, have undergone some kind of initiation.

Attempts were made to interpret lakshana as referring to the wearing of the sacred thread (yajnopavita) by Brahmins and the branding of shoulder blades in the case of Sundras and women. This is untenable as the word 'lakshana' which qualifies jointly and simultaneously four or five groups of persons, is made to bear one meaning when applied to one to one group and another meaning when applied to other groups. Though the Kalpasturas do not refer to the sacrament of tapa, there is Vedic sanction for Brahmins being marked by heated metal with the forms of the conch and the discus on their arms. Though the sacrament of pancha-samskara makes men and women of all castes qualified to perform formal worship of the icons of God, not every one can do so in the temple. The right is restricted to those who get initiation. At home also, not all perform the worship: it is usually done by the eldest male member of the family. Though women are eligible, normally they do not do it, in spite of tantrika mantras being available to them in the place of Vedic mantras.

Seventy-four preceptors or acharya-purusat, also called simhasanadhipatis or "occupants of the (spiritual) throne", are said to have been entrusted by Ramanuja with the duty of administrating the sacrament of pancha-samskara. The Yatiraja-vaibhava (103) and the Guruparamparas refer to this. But there does not seem to be any written evidence to show that the right to give the fivefold initiation is restricted only to the 74 and their successors. As of today, only a few of the successors exercise this privilege, other having given it up for some reasons or other. The ritual of initiation is begun by the acharya himself with a homa in the sacred fire. The Purusa-sukta, Sri-sukta and other sacred texts are recited. Metallic representations of the conch and discus are 'bathed' ceremonially and then heated in the fire. They are then applied by the acharya to the shoulder blades of the disciple, who also is given the pundra. The teaching of three mantras – the mula mantra (*i.e.*, the astaksara), the dvaya and the caramasloka (*i.e.*, Gita XVIII. 66)-constitutes the samskara of mantra. As for ijya, the disciple is instructed in the ritual of worshipping God at home. He is then given a name by which he can regard himself as a devotee of God and His devotees, 1458. The occasion is utilized also to teach the disciple the significance of the epics, the Gita and some other sacred texts and instruct him on how to lead his life as a Srivaisnava.

The acharyas include some ascetics. This has given rise to a practical difficulty. Sannyasins are prohibited from offering obligations in the fire. Hence the homa required for the sacrament has to be performed by some grihastha attached to the sannyasin who however applies the heated metal pieces to the shoulder blades of the initiates. For some time past, the tendency has been asked to seek initiation at the hands of the sanyasins presiding over mutts, and the disciples of acharya-purusyas who are householders, are dwindling. One group of Vaishnavas, it may be noted, do not receive pancha-samskara. The Vaikanasa claim that they are children of the Lord and that they do not need sacrament or preceptors.

Other Vaishnavas are distinguished from non-Vaishnavas mentally, verbally and physically. They engage in the distinctive mental activity of bhutasuddhi and the verbal activity of japa in the chakrabjamandala and are subjected to being marked on the body by the metallic emblems of the conch and discus on the Lord. The Vaikhanasas reject the three mantras taught to other Vaishnavas by their preceptors. Japa in the chakrabjamamdala is meant only for those who follow the

Pancharatra exclusively and without reference to the Kalpasutras. As for tapa, the Vaikhanas claim that in their families the fetus in the womb gets the marks of the conch and the discus in the eighth month of the pregnancy of the mother when the ritual of Vishnubali is performed. Narayana Himself is said to give the mark to the fetus. Thus the followers of the of the Vaikhanasa-sutras claim to be Garbha-Vaishnavas, that is, Vaishnavas even in the womb of their mothers.

In spite of these differences between the two Agamas, it may be noted that, according to both, the Vaishnava has to bear the marks of the conch and discus. It should also be remembered that this samskara has for long been considered an essential qualification for conducting worship and preparing food both at home and in temples. Even women must be thus qualified before they can handle household vessels, bring water or cook and server food. The pious would not accept even a cup of water from one who had not undergone the samskara though the strict observance of this ritual is no longer common. The daily routine has been prescribed for the Vaishnava who has undergone pancha-samskara. An immediate disciple of Ramanuja-known as Vangi Vamsesvara and Sriranga-narayana-earya – has a written manual, Ahnika-karika, based on panchakala-prakriya, 1460.

It should be deemed an authentic account of Ramanuja's views. The round of daily religious exercises differs in some essential particulars in the Pancharatra and Vedic traditions. The latter enjoins the Gayatri-japa, 1461, thrice daily. The Pancharatra prescribed the japa of Vishnu-mantras. During japa, Vishnu is to be mediated as being present in the sun's disc, and He is to be worshipped with specific names. The mantra for the japa in the morning is Vishnu-Gayatri, the devata being of Vishnu; at midday, it is dvadasaksara invoking Maha-Vishnu; and in the evening, it is astaksara, the Deity being Sada-Vishnu, 1462.

Vedanta Desika warns against hastily discarding Vedic practices and adopting in their place directions from Pancharatra texts except on very strong grounds. He points out that followers of Ramanuja's teachings have performed marriages, upanyana and other sacraments and rituals in accordance with the Sutras to which they owe allegiance. Further, all know that Ramanuja's preceptors like Periya Nambi performed sacrifices and other rituals according to their Sutras. Hence those intent on following the precepts and practices of Ramanuja and his disciples should not adopt for rituals and ceremonies the mantras enjoined in particular Pancharatra samhitas, 1463.

The Ahnika-karika enjoins homa and other rituals of the kind mentioned in Kalpa-sutras, 1464. In this context, the question arises as to accepting the Pancharatra injunctions about rituals involving the slaughter of animals. King Uparicara Vashu is said to have offered as victim in a sacrifice the figure of an animal made of the flour of wild grains, 1465. Brihaspati, the preceptor of the gods, accepted this view after initial opposition, 1466.

The Pancharatra is held to be Vashu's authority. The Dvaitins, influenced by the humane considerateness of the Pancharatra, recommended sacrifices of this kind, 1467. The Visistadvaitins, though far more influenced by the Pancharatra, argue that a Vedic ritual, if obligatory or required on specific occasions, must not be given up or modified, 1468. The offering of arghya to the sun during sandhya prayers is cited as a ritual involving himsa, as it brings about the death of the demons impending the progress of the sun, 1469. One has to turn around oneself in a clockwise direction to get rid of the sin of killing. But the killing of an animal as a victim in a sacrifice is not deemed a sin. Himsa or causing injury is an act which leads to evil. In this sense, there is no himsa in sacrifices, as the victims attain to a better state after death. (Sribhasya III. 1.25)

Therefore the ritual of killing animals is meritorious. It inflicts beneficial injury like surgery. (Cf. Adhikaranasaravali, 284-6, of Vedanta Desika on ibid.) As regards the japa of any mantra, it can be verbal, silent or mental, securing protection from evil spirits, accomplishing desired objectives and helping in the attainment of moksha respectively, 1471. The rosary, aksa-sutra, is to be used for counting the number of repetitions, 1472. Interruptions to japa are not generally permissible, but the arrival of a devotee of the Lord requires a break in it, 1473. Devotion to the devotees of God is stressed so much. Besides, japa, four other rites are enjoined by the Pancharatra– “sacrifice”, liberation, ritual bath and offering food to Brahmins. Collectively, the five are known as purascharana, 1474. They are needed for securing full spiritual power through mantras.

Among the prescribed mantras, are Vishnu-Gayatri, Bhuta-suddhi mantras, Shakti-mantra, Panchopanisan mantra, Vaishnava mantras generally and many others. Of these, only a few are used in japa and for attaining specific objectives. The Savitri is obligatory for everyone qualified to study the Vedas. Among the Vaishnava mantras, the pranava, sadaskara, astaskara, dvadasaksara and Jintante are described as vyapaka. The ashtakshara is universally enjoined by Vaishnava tradition in addition to the Gayatri, 1475.

Worship of God is obligatory for everyone. It is of two kinds,-in temple and at home. The former known as prarthna-yajana (worship for the sake of others) is concerned with the consecrated idols in the temple and intended for the welfare of the community. The Agamas however suggest that ‘paratha’ means ‘concerned with what is superior’ or ‘excellent’. It is like the sun, while worship at home is like a domestic lamp. It is also declared that parartha, 1476 worship leads to final liberation.

Poigai Alvar

Poigai Alvar was one of the twelve Alvars and was a Hindu saint. He belonged to the Vaishnavite (worshipper of Vishnu) faith. Poigai Alvar was born in a village near Kanchipuram in the seventh century CE. Poigai (pond in Tamil) was found as a baby near a pond of a Vishnu temple. Vaishnavites consider him to be an incarnation of *Panchajanyam*, the divine conch of Vishnu. Poigai Alvar’s 100 hymns form the beginning of the Naalayira Divyap Prabhandham. Poigai Alwar’s hymns start with the words *Vaiyam Tagaliya, Varkadale Neyyaaga* (I am singing these garland of verses (pasurams) and dedicating to Him, whose vision I had is the light of the lamp of the earth, and oil being the seas, the sun being the source of the light).

Bhoothathalvar

Bhoothathalvar was one of the twelve Alvars and was a Hindu saint. He belonged to the Vaishnavite (worshipper of Vishnu) faith.

Bhoothathalvar was born in the Pallava country near Mamallapuram in the seventh century CE. Bhoothathalvar's 100 hymns form the second part of the Naalayira Divyap Prabhandham. Bhoothathalvar's hymns start with the words *Anbe Tagaliya, Aarvame Neyyaaga* (with love as the lamp and devotion as the oil). He is considered an incarnation of Vishnu's mace, Kaumodaki.

Anbe Thagliyaa Aarvame Neyyaaga
Inburugu Chintai Idu Thiriyaa
Nanpurugi Gnaana Chudar Vilakku Etrinen
Naaranarku Gnaana Thamizh Purindha Naan

"I who wrote this song that bestows wisdom, with love as the lamp, endearing involvement as the lubricant ghee,

and knowledge as the wick of the burning torch, dedicated myself to the service of the Lord"

Peyalvar

Payalvar was one of the twelve Alvars and was a Hindu saint. He belonged to the Vaishnavite (worshipper of Vishnu) faith.

Payalvar was born in Tirumayilai (Mylapore, now part of he city of Chennai om the seventh century CE. Vaishnavite traditions states that Peyalvar was found on a lily flower in the pond of the Adi Keshava Perumal Temple in Mylapore. They also believe that he was an incarnation of the sacred sword of Vishnu, Nandaki. Payalvar's 100 hymns in the Naalayira Divyap Prabhandham are the third 100 and start with the words *Tiruk Kanden, Pon Meni Kanden* ('I found the glorious, golden form of the Lord').

Tiruk Kanden Pon Meni Kanden-Thigazhum
Arukkan Ani Niramum Kanden-Seruk Kilarum
Pon Aazhi Kanden Puri Sangam Kai Kanden
En Aazhi Vannan Paal Inru.

"On witnessing the glorious vision in which the entire universe was the very body of Lord Narayana, the Alwar proceeds to describe all that he had seen."

Peyalvar's hymns are of the *Anthathi* style, in which the last word of the previous hymn is used as the first work in the next hymn.

Thirumalisai Alwar

A tamil mendicant-saint revered in the srivaishnavismschool of south India probably between the 4th to 8th century CE. There are no historical records of this alvars. The legend of this saint devotees of Sri Vaishnavism say that he was the incarnation of Vishnu's disc, Sudarshana. He may have been born a Brahmin but was adopted by a lower caste person, as he proclaimed that he was "not of the 4 castes" in one of his couplets. He seems to have had the access and resources to learn about various religions because in one of his couplets he says "After my futile efforts to know the supreme principle through jainism, buddhism, and saivic philosophy, I am blessed by sriman Narayana to take refuge at the Lotus feet of Sri Devi and have escaped all problems and misfortunes since then".

There is also evidence in his couplets that show that he was discriminated against by brahmins and asked God to help reveal their ignorance to them. Legend also says that when he displeased the pallava king and was banished, he asked the God from the temple, Yathotkari, to leave with him. Vishnu is said to have rolled up the snake Sesha like a matress and left with him. There are only 216 of his parasurams in the 4000 Divya Prabhandham.

Nammalvar

Nammalvar (also Nammaazhvaar, Nammazhvar, Nammaalvaar) was one of the twelve Alvars, well-known for his many hymns on devotion to Vishnu. Tradition gives him the date 3102 B.C. (*i.e.*, the beginning of the kali yuga), but scholars give him a date 880–930 A.D which is more realistic based on the events recorded. He was born in the asterism Visakham, in what is now Alvartirunakari (also known as

Tirukkurugur), Tamil Nadu. His name means "our own alwar" (alwar means "one immersed in God"). He was also known as Maran and Catakopan.

Child Prodigy Discovered

It is better to make a distinction between traditional scholars and academic, or critical, scholars. The academy does not own the term scholar, which does carry some prestige. In the politics of representation, traditional scholars want to be identified as such, not excluded from "scholarship".

He must have been born fully enlightened because as a baby he never cried or suckled and never opened his eyes. According to tradition, as a child he responded to no external stimuli and his parents left him at the feet of the statue of Vishnu. The child then got up and climbed into a hole in a tamarind, sat in the lotus position, and began to meditate.

It appears he was in this state for as long as sixteen years when a Tamil poet and scholar in North India named Maturakavi saw a bright light shining to the south, and followed it until he reached Nammalvar's tree. Unable to elicit any reaction from the child, he asked him a riddle: "If the small is born in a dead's body (or stomach), what will it eat and where will it stay?" meaning, if the subtle soul is embodied in the gross body, what are its actions and thoughts? Nammalvar broke his lifelong silence and responded, "That it will eat, and there it will rest!" meaning that if the soul identifies with the body, it will be the body but if it merges with the divine, it will stay in vaikuntha and eat (think) of God. Madhura-kavi realized the holiness of this child yogi.

First Disciple

Madhura-kavi was himself in a state of high spiritual evolution when he asked Nammalvar the "right" question and made him speak. Immediately he took him for his Acharya (Teacher, Guru). Nammalvar consented to being his guru, instructed him in the secret doctrines of Vaishnavism and thereafter could not resist the urge of singing about God and trumpeting about Him. He composed on the spot a thousand hymns praising Vishnu, each one starting with the last word of the previous poem. We owe it to Madhurakavi for setting them to music. Madhurakavi became the boy's student and went on to compose poems about his prodigal master.

Writings

His contribution of four works (numbering 1298 hymns) to the four thousand of the Divya Prabhandham includes the entire fourth thousand and part of the third thousand these works are

- Tiruvaymozhi (1102 verses),
- Tiruviruttam (100 verses),
- Tiruvasiriam (7 verses) and
- Periya Tiruvanthadi (87 verses).

Tiruvaymozhi describes ranganatha as a metaphor to discussing the philosophical details in

- the nature of the paramatma
- the nature of the jeevatma
- the means for the jeevatma to attain the goal of Paramatma
- the blocks and hurdles on the way and
- the goal moksha.

In the Sri Vaishnava canon these four represent in the Tamil language the four Sanskrit vedas, respectively, the Sama Veda, Rig Veda, Yajur Veda and Atharva Veda. According to tradition "He poured the cream of these vedas" into his songs and poetry that were the result of deep mystic experience. Though Nammalvar did not visit any of the 108 divyadesam temples talked about in the Vaishnava religion it appears from his works he must have had the vision of all the archna forms in the temples he had glorified in his hymns.

Acme of Devotion

The subject matter of the four works was certainly the five principles, namely, the Lord, the soul, the means, the end, and the obstacles to spirituality. Through all this ran a thread of the acme of devotion to Lord Krishna. Whatever Krishna ate, whatever He drank, whatever betel He chewed was the dearest to the saint. The supreme object of life was to be at the Lord's lotus feet and to serve him eternally in blissful love. Therefore seek Him all your life, praise Him, surrender to Him, speak of His glories and exploits, revel in His majesty and continue the recitation of His names. This was his message to the world at large. Like the Buddha who appeared in the northernmost part of India and finally engulfed not only India but Asia and the whole world by his teachings, Nammalvar was the star of the southernmost part of India whose work engulfed the whole world of Vaishnavism.

A Sample Hymn

Makes me serve him all the time and everywhere;
Anchors me thus by seating himself in my mind;
Swerves me not by gracing me in his service alone;
Krishṇa's wish is my wish, my pleasure, my glory.

(3102 of Sacred Book-By Sri. Srirama Bharathi Araiyar)

Periyalvar

He was born in a Brahmin family in Srivilliputhur, near madurai in the 6th to 9th century CE and was named vishnuchittar meaning "one who has Vishnu in his mind". Legend says he rejected the vedic philosophical debates of his upbringing to focus bhakti especially on doing simple tasks for God. He would make garlands of flowers for the deity of the temple. One day the pallava king of the land had a competition between scholars to find one who would explain philosophical principles. Vishnuchittar won the competition by explaining that the path to moksha is by service to God. Legend goes that the king honoured vishnuchittar and God himself came down to earth to see this. Vishnuchittar composed a couplet called "pallandu" (also called *thirup-palandu* out of respect) translating to "Long live for many years, Long live for many years for Hundreds of thousands of years!" to God.

This is a very important prayer in Sri Vaishnava liturgy today. Vishnuchittar composed some pasurams in the 4000 Divya Prabhandham called perialvar tirumozhi where he explores a devotee's love for God through the metaphor of Yashoda's motherly love for Krishna. He was the adopted father of godha or Andal, the only woman alwar.

Madhurakavi Alwar

A Tamil saint/composer of the 6th to 8th century. He composed 11 couplets in praise of his guru, nammalvar in the 4000 divya Prabhandham. One of the alvars He is venerated as the dawn Ushas before the sun rise of nammalvar as he was born before nammalvar. There is no historical record of his life but legends say that he was a well read and well travelled Brahmin with a gift for music. He set nammalvar's compositions to music.

Aandaal

Aandaal is a 10th century Tamil saint and one of the twelve Alvars (saints) and the only woman Alwar of Vaishnavism. She is credited with the great Tamil works of Thirupavai and Nachiar Tirumozhi that are still recited by devotees during the Winter festival season of Margazhi. Aandaal is known for her unwavering devotion to Lord Vishnu and the Hindu temple at Srivilliputhur is dedicated to her. Adopted by her father, the famous saint Periyalvar who found her as a baby, Aandaal avoided earthly marriage, the normal and expected path for women of her culture, to "marry" Lord Vishnu, both spiritually and physically. In many places in India, particularly in Tamilnadu, Aandaal is treated more than a saint and as a form of God herself, similar to tradition of Lord Buddha and Jesus Christ. The Vaishnavite devotees thus consider her to be a form of Bhuma Devi (Mother Earth), the divine wife of Lord Vishnu.

Kulasekaraazhvaar

Born in the asterism Punarvasu as Kulasekara Verma, he was king of the Chera Kingdom and lived in Kollam in Kerala. After ruling for a few years, he gave up his throne, became a sanyasin and is revered as the 9th alvars (one of 12 mendicant saints venerated by south Indian Sri Vaishnavism) and wrote bhakti songs filled with yearning towards God called

Parasurams which are an important part of carnatic music classical repertory.

A great devotee of Rama, he considered the painful experiences of Lord Rama to be his own. He is therefore also known as 'Perumal', meaning 'The Great' – which is usually the epithet for the Lord. His devotion was so intense that he worshipped the devotees of the Lord as the Lord Himself. He lived in Srirangam and was serving the deity Ranganatha in the temple there.

Works

- mukundamala
- Perumal Thirumozhi (Lord's sacred poem) containing 105 verses in Tamil, many of which are dedicated to Lord Rama and Lord Krishna and Ranganatha, a part of the 4000 Divya Prabhandham.

Thondaradippodi Alwar

Thondaradippodi Alwar, who is also named as "Vipra Narayanar" led his life as the slave to the Perumal and worked for the Perumal by dedicating him with Garlands. He is one of the 12 Alvars.

Birth and Early Life

Thondaradippodi Alwar was born in a small village by name 'Thiru mandaggudi' in Prabhava year, Margazhi month, Krishna Chaturthi, Tuesday in Kettai Nakshatram (star). His father 'Veda Visaradhar' belonged to "Kudumi Sozhiyap Brahmanar" community also called as "Vipra" people, whose routine work is to praise about Sri Vishnu.

On the 12th day after his birth, he was named as "Vipra Narayanar". He was properly taken care of and proper education along with the bhakti towards Sri Vishnu was teached to him. He grew up with all good characters and had a good structured body, too. In spite of being good and beautiful and dedicated to Sri Vishnu bhakti, he treated all the aged persons and persons who are younger to him in the same way and gave proper respect to them.

Reaching Srirangam

One day, "Seenai Mudaliyaar", who was the leader of Sri Vishnu's troop came down to the earth and explained to Vipranarayanar, why he was born in this mighty world and what are the things that he should follow to attain Mukthi.

On hearing the truth of his birth, he showed much love, affection and bhakti towards Sriman Narayanan. As an extension to it, he started thinking what should be the next step to be taken to explore his bhakti. He thought of travelling to all of the temples where the Emperumaan is giving his seva. But, he had a question that from which sthalam he should start with. After a long thinking, he finally decided that he should start from the greatest of all of the Vishnu sthalams, the Srirangam, which is located on the banks of cauvery river should be the starting sthalam he should worship and started travelling towards it.

After reaching the temple, he went inside the moolavar sannadhi and was very happy to see the perumal on the Aadhiseshan (the snake) who is found in sleeping posture. After seeing the beauty of Sri Ranganathar, he did not want to see any of the things around and thought his bhakti and the remaining life should be spent completely for Aranganathan. He thought in what way he can express his bhakti towards the emperumaan and finally he thought he can offer him with the Garland daily to the perumal.

Soon after this thought, he constructed a big Nandhavanam (flower park) in Srirangam, where various beautiful and fragrance flower plants are grown. He built a small hut in the midst of the park. Daily, in the morning, as a daily work, he dedicated the Garland to Sri Ranganathar and only then did his routine jobs. Since, he thought, only Sri Aranganathar is the beautiest thing in this world, he didn't turn up towards any of the women around him. Because of this, all the people thought he doesn't have any feelings towards the people around him. On seeing these kinds of scenes from Sri Vaikundam, thirumagal, Sri Lakshmi questioned Sriman Narayanan that because of having the affection towards him (Sri Ranganathar) and in spite of being a male, he doesn't know how the love should be expressed towards a women and continued that which women can show her love towards a person like Vipra Narayanan? Knowing the future and what is going to happen, Sri Vishnu casually answered Sri Lakshmi that she is going to see what is going to happen for him.

Devi and Devadevi

In thirukkarambanoor, there lived two Daasis (Women who dances in front of the king to earn money) by named Devi, who is the elder sister and Devadevi, who is the younger sister. They wanted to dance to the Urayoor king and to get prize from him. Then both of them went to Sri Rangam temple to worship Aranganathar. When they were going to the temple, they had the chance to see the Nandhavanam, which belonged to Vipra Narayanar. Devadevi wanted to see him and they both went to the small hut. Devadevi, on seeing him, was very much attracted towards him and started loving at the first sight. She expressed her love in words and continued with his work. She did service to convert him towards her side. Soon she completely turned Vipra Narayanar towards her and they both started to live together.

One day, Devadevi said to Vipra Narayanar that she wanted to see her parents and her sister. Vipra Narayanar said that he would too come along with her, since he could not live without leaving her for a second. But her mother did not want him to be in the house and asked Vipra Narayanar to get out of the house for he did not have any wealth. Since, Vipra Narayanar is helpless, he came out of the house and entered a Nandavanam and being so tired, he slept for a while. On seeing this, the Emperumaan wanted him to get of the Maya and wanted him to get along with the right path, which leads to the Thiruvadi of Him (Sri Vishnu). He changed himself as a Student of Vipra Narayanar and went straight to Devadevi's house and gave him a big silver vessel as given by his Guru to Devadevi.

On getting that her mother thanked him and asked his Guru, Vipra Narayanar to return back to her house. On hearing this, Vipra Naryanar walked fast to the house and hugged

Devadevi. The Next morning, the archakar entered into the Sri Ranganathar sannadhi and found one Vattal (Vessel which is used to do the daily pooja) was missing and complained to the king about this. The king ordered his servants to go in search of the Vattal and they found that in Devadevi's house.

Vipra Narayanar was jailed for one night. On that night, Vipra Narayanar cried and asked the Emperumaan why like this has happened to him and prayed towards him. On the same night, the Emperumaan arise in the dream of the king and explained him what had happened and asked him to free Vipra Narayanar. The king explained that this is a small thiru Vilayaadal (game) of Sri Emperumaan.

Knowing the real truth why he came to the world, he went to Sri Ranganathar temple and praised the Lord. Then, he worshipped all the bhaktas of Sri Vishnu and put the podi (Small tiny dust particles), which is found under the feet of them in his head and sang songs in praise of Sri Ranganathar.

From then, he was called as "Thondaradipodi Alwar". Finally he realised for what he came for to the earth and went to all the Vishnu temples to praise him.

Literary Works

He composed Tirumaalai (45 paasuram) Tiruppalli Ezuchi (10 paasuram)

Alvars

The Alvars were twelve Vaishnava saints, who during the early medieval period of Tamil history (between the seventh and the tenth centuries CE) helped revive devotional Hinduism (bhakti) through their hymns of worship to Vishnu and his avatars. They included a woman (Andal). The collection of their hymns are known as Divya Prabhandham

The Bhakti literature that sprang from these Alvars has contributed to the establishment and sustenance of a culture that broke away from the ritual-oriented Vedic religion and rooted itself in devotion as the only path for salvation. In addition they helped to make the Tamil religious life independent of a knowledge of Sanskrit.

Etymology

Alvars or 'Azhwars' literally means 'people who are immersed'. They are so called because they were immersed in their devotion and love to their Lord, Vishnu.

Legacy

The twelve Alvars were all inspired and ardent devotees who transmitted their divine infatuation to millions. They have left behind an imperishable legacy of devotional Tamil poetry-naalaayira Divya Prabhandham (considered to be the essence of the Vedas, in Tamil, and all in praise of Lord Vishnu). These have been rarely equalled either in quantity or in quality ever after. The one held in greatest esteem among the Alvars is Nammalvar. He lived during the seventh century CE. His contribution to the four thousand prabandhams is as many as 1352. His hymns are considered by the Vaishnavites to contain the essence of the Vedas. His works-*Thiru Aasiriyam*, *Thiru Virudham*, *Periya Thiruvandhadhi* correspond to the Yajur, Rig and Atarvana veda respectively. His other work *Periya Thirumozhi* (Divine words) is the one of the key works of Vaishnavism.

Periyalvar delighted in worshipping Vishnu as mother, nurse, devotee and lady love. Andal, who grew up in Periyalvar's home, is attributed the *Tiruppaavai*, a most beautiful collection of 30 verses giving expression to the purest love of God.

The Twelve Alvars: Bhaktisara

There were twelve Alvars who appeared in South India. Not all at the same time, but over a period of several centuries. They established the basis of the Krishna bhakti cult in the Kali-yuga. The appearance of such great devotees in the Kali-yuga is predicted in the Srimad-Bhagavatam. Srimad-Bhagavatam was spoken at the beginning of the Kali-yuga, and when Krishna left this planet then he took with Him dharma. The Vedic dharma at that point disappeared, or became invalid, and spiritual knowledge was also obscured.

But it says in the same verse that Lord Krishna left the Srimad Bhagavatam for the people in Kali-yuga to get light out of. Now still, the book Bhagavata was there but they also needed the person Bhagavata, or one who lives the Srimad-Bhagavatam. In other words, they needed the spiritual master. So in the initial stage of Kali-yuga, the first few centuries, these twelve Alvars appeared in South India, and actually established the basis of what would later on become the four Vaishnava sampradayas. The four sampradayas all had their origin in South India, and the founders of these sampradayas each in their own way drew, to a greater or lesser extent, from this tradition of the Alvars, especially in the Lakshmi sampradaya, but it is also there in our sampradaya too, and in the others.

The Radha Krishna cult is the further development of the devotional tendency of loving God in close fellowship and in the spirit and relation of a woman to her husband or lover. This tendency is striking in the Prabandham of the Alvars. Goda, the famous woman Alwar, is said to have been married to the Deity Ranganatha of the Sri Rangam temple. Tondaredipodi Alwar (Bhaktanghri Renu in Sanskrit) expresses in his Tirup Palliyeducci (Paramatmar Jagarana in Sanskrit) that to serve and love God one's spiritual body is the summom bonum of one's service to God. One of the Alvars was known as Bhaktisara. He was living at Sri Rangam, and he was living very, very simply, like a babaji. He would sit at Sri Rangam. Around the temple are 7 walls, and between the walls are some areas like streets where people would walk and live. So Bhaktisara would sit in that street area, or in the courtyard of the temple, and he would sit in the sun and sew his cloth. He wore old, old cloth, and he would sew it and maintain it until practically speaking it was unwearable, and only then he would he go and find some new old cloth.

And because he purposely kept himself impoverished, he used to sew it with a broom-straw, this would be his needle. And on the end of it he would tie a thread. Then he would take his old torn garment and sew it. So he used to be seen often, sitting somewhere in the Sri Rangam complex sewing his cloth. One day Lord Shiva and Parvati were flying over the sky just above the earth.

By flying like this, their shadow was cast upon the ground. So where ever this shadow would fall, naturally people became very enlivened in their material desires. But when the shadow fell on Bhaktisara, they stopped and observed him. Bhaktisara was sewing, and then he looked around and saw this shadow. He could understand that this shadow is inspiring one to have material desires. So thinking, "This is inauspicious," he moved out of the shadow.

Parvati turned to Lord Shiva and said, "What is this? Everyone else is so eager that we may give benediction, that they rejoice if they catch sight of us or if our shadow falls on them. But this person is moving out of the shadow. What does this mean?"

Lord Shiva replied, "He is a great Vishnu bhakta. We have no business with him. He doesn't need anything from us."

Parvati said, "No, no, no. Everyone wants something from us. Everybody has material desires and wants them fulfilled. Let us go down there."

So they went down there, Parvati and Shiva, and they appeared before Bhaktisara, who was just sewing his cloth. Then Parvati, who was standing behind Lord Shiva said, "Offer him heaven."

Lord Shiva said, "My dear devotee of the Lord, we have come down before you to give you a benediction, so I am offering you the svarga-loka, the heavenly world. You may go there."

Bhaktisara didn't even look up, he just kept on sewing and ignored them. So then Parvati said, "Offer him the post of Lord Brahma."

Bhaktisara ignored this also, and continued sewing his cloth. Parvati was becoming more and more agitated, so she said, "Offer him a form just like yourself."

Lord Shiva said, "So will you take a form or a position just like mine?"

Then Bhaktisara looked up and said, "You've already got enough trouble." He was referring to the fact that although he was the great Lord Shiva, he was still being pushed around by Parvati.

So then Parvati said, "He must ask something from us. So just tell him to ask for a benediction."

Lord Shiva said, "All right, if you don't want any of these things I am offering, then you pick something, because you must take a benediction from us."

Bhaktisara said, "Well, you give me liberation from material existence."

Lord Shiva said, "That's one thing I can't give." Lord Shiva himself says "mukti pradata sarvesam Vishnu eva na samsaya." One who wants mukti has to approach Lord Vishnu. So he was caught in an embarrassing situation when Bhaktisara asked for that, something he could not give. "I'm sorry, but this is something that I cannot give."

Bhaktisara said, "Then what is the use? What is the use of you and your benedictions?"

Lord Shiva said, "You must take something from us, because we have come to offer you a benediction."

"All right," Bhaktisara said. "While I sew, the thread is always coming off the needle, because that is only a broomstraw. So you please make some arrangement that this thread becomes attached to the back of the straw so I can finish sewing my cloth."

At this, Parvati became enraged. "He's insulting you! You can give him the universe, and he's asking for such a minuscule thing. This is an insult. You must do something."

So then Lord Shiva opened his third eye, and a tongue of flame came out towards Bhaktisara. Bhaktisara saw this tongue of flame coming, so he pressed his foot to the ground, and from where he pressed his foot an even greater flame came out, devoured that flame from Lord Shiva's eye, and started to pursue Lord Shiva. Lord Shiva was running around Sri Rangam temple with this fire following him. So he went before the deity of Ranganatha and prayed, "My dear Lord, I have made a great mistake today by offending your devotee. Please help me." So the Lord instantly conjured a Vaikuntha cloud, and this was raining a kind of nectar. Because it was a Vaikuntha cloud, within a very short time the whole area of Sri Rangam was totally flooded. In this way the fire of Bhaktisara was extinguished, but the whole temple was covered, like a big lake.

Then suddenly Bhaktisara appeared on top of the lake, like a cork. He was just sitting on top of the water still sewing his cloth. Very quickly the water subsided. Lord Shiva appeared before Bhaktisara and blessed him. He gave him this name-Bhaktisara. Bhakti means devotion, and sara means essence. So this name means, "the essence of devotion." Lord Shiva said, "Your attitude and devotion to Lord Vishnu is the essence of devotion, because even if Lord Shiva comes, you do not want anything from him, although all the materialists are desiring."

Another time there was a shuktihara, a magician flying through the sky on his tiger. Bhaktisara was again sitting sewing his cloth. And then when he flew over Bhaktisara, his tiger couldn't stay in the sky, it fell to the ground. He came down to earth, and he was wondering, "What is this? Why can't we fly here?" And then he saw Bhaktisara, and he could understand, "Oh, it must be by the influence of this saintly person that my magic is rendered null and void." The magician became pleased, because he could understand "He is very, very advanced. He is so pure that this kind of mystic jugglery I am doing does not function in his presence." So being very pleased with him, he wanted to reward him. He took from his shoulders a very, very expensive ornamental cloth that he was wearing. You can just imagine, he's a wizard so he's got to be dressed really far out.

It had gold sewn into it and diamonds and rubies and emeralds all over it. So he took it off, and he presented it to Bhaktisara, "You please accept this." As soon as the cloth touched Bhaktisara's hand, it was transformed into an old rag with holes, the type which Bhaktisara was accustomed to wear, something fit for Bhaktisara to wear. And then Bhaktisara took his cloth that he had been wearing, and gave it to this shuktihara, and as soon as the shuktihara took it, the cloth became like molten diamonds, like a diamond jelly. It was very valuable, and the shuktihara was very pleased to receive this.

5

Belief and Concept in Vaishnavism

Belief and Concepts

Introduces the beliefs, ideals and traditions of the Hindus, a heterogeneous people, who can trace their faith back to a civilization that existed over 4,000 years ago.

The beliefs and customs that India mystica deals with are those encountered in everyday life. We feel the need to delve into such a subjective and potentially controversial area, Because understanding the How, What and why of many customs and practices can only happen with reasoned information.

Hinduism is a blanket term for a staggering range of concepts and 'levels' of thought. It is also a pulsating record of the cultural evolution ρf 80 percent of the world's largest democracy and the millions-strong Hindu Diaspora that pervades almost all continents. Many Hindu practices and beliefs astound with their scientific and humane relevance.

Others seem superstitious and illogical. Nevertheless, it is interesting to understand why those who went before us acted in a particular-often apparently contradictory-way. Flowers, fasts, feasts, adornment, rites of passage, sacred scriptures, deeply felt environmental and social values. Beliefs and Concepts that are generally followed but do not pertain to a particular ceremony or time.

Shringaar

Everybody likes ornaments to adorn themselves. Especially ladies like adorning each part of the body. There is a tradition in most of the part of the world. Even Adiasis also have their own ornaments made from wood, bones, stones and things. Traditional a married Indian woman had in her cachet a wealth of adornments. Passed down to her through generations were the formalized yet unwritten laws of solah singar, sixteen adornments or basic make up norms.

The meaning of bhushaṇa is to adorn and Abhushan means the ornaments or things, which is used to adorn. Where health maintain the physical beauty, Abhusana enhances it.

There is a special scientific reason to wear ornaments. Each has their special effect on the body and mental & physical health. The material does matter for ornaments because they are made from different material like gold, silver, platinum etc. some times pearls, diamonds, gems stones are attached to it. Mainly gold and silver are used as a row material to making ornaments because the effect of silver is cold and gold hot. Both metals reduce the chances of cancer and balance the blood circulation.

Astrology says that there is a special influence for every ornament for every people. Not only that it is also important that where it to wear. According to them the effect of planets has great influence to person's mental, physical health and environment.

Bindi

Bindi is derived from bindu, the Sanskrit word for dot. It is usually a red dot made with vermilion powder, which is worn by women between their eyebrows on their forehead. Considered a symbol of Parvati, a bindi signifies female energy and is believed to protect women and their husbands. Traditionally a symbol of marriage (vivaha), it has also become decorative and is worn today by unmarried girls and women as well. Widows however, usually do not wear a bindi.

No longer restricted in colour or shape, bindis are seen in many bright colours and in different shapes and designs. They are also made of coloured felt and embellished with coloured glass or glitter.

Kaajal

This is a thick, black, ointment, made of ground lead sulfide or antimony sulfide, which is used as an eye liner, a coolant, and protector against the evil eye.

Kajal is usually applied with a fingertip on the inner rim of the lower eyelid and sometimes on the upper lid as well. Though traditionally used by both men and women, today largely women use kajal. However children of both sexes are made to use kajal as a coolant and to protect them from evil. Indeed Indians of all ages and faiths use it for these reasons.

From the time a child is six days old, its mother applies kajal to its eyes and also a small black dot on the forehead to mar the child's beauty. This 'imperfection' is said to protect from evil. Kajal can be made at home by mixing the soot of an oil lamp with clarified butter, while commercially it is available in small boxes or as pencils.

Flowers

Flowers are considered symbolic of life and happiness and are therefore an important aspect of any worship. Each deity is believed to have a favourite flower, which, if offered in worship, adds merit to the prayer. Usually, fragrant soft-petalled flowers are offered in worship. Red flowers are usually offered to female deities.

Traditionally, showering honoured people with flower petals was an accepted form of demonstrating respect. Kings, heroes and sages were honoured in this manner. Even today, a traditional welcome or farewell is incomplete without flowers.

This practice is also evident in the course of the wedding ceremony (see Vivaha). Showering petals is also a form of blessing, as at the end of the ceremony, when all those present shower the Flowers bride and groom with petals. In the southern part of India, an important part of a woman's toilette is flowers in her hair.

Mehendi

Mehendi or henna (Lawsonia inermis) is a small tropical shrub. Its leaves and shoots, when ground into a paste, give out a rust-coloured pigment. This is used as a hairdye and as a cosmetic to make intricate designs on the palms and feet. Apart from its colour, henna also has a pleasing fragrance and a cooling effect.

Henna's auspicious reddish colour merits its use during all ceremonies related to marriage and married women. It denotes the prosperity that a bride is expected to bring to her new house and is a part of the 16 traditional adornments, with a special ceremony of its own, simply called Mehendi. A day before the wedding, all the friends and relatives of the bride gather together, and the dark green paste of henna leaves is applied in decorative designs to the bride's palms and feet. The paste can be made by grinding fresh leaves, or by powdering dried leaves and then mixing the powder with water. A professional is usually called upon, for the

Process is almost an art form. The henna paste is squeezed through the tip of a cone and by systematically moving the cone, the designs are 'drawn'.

After this, the paste must be allowed to dry until it becomes hard and crusted. This takes three to four hours. During this time, the bride must sit still. She gets some much-needed rest and listens to the advice of friends and relatives who are also participating in this ceremony. Because of its cooling effect, henna also calms the bride's nerves. After it dries, it is washed off or scraped off with the help of a blunt knife, leaving a dark rust imprint, which stays for about a week. After the wedding, the bride is not expected to do any domestic work till her henna fades.

Apart from weddings, henna is also applied during the various vratas observed by married women. It is very popular with both men and women as a conditioner and dye for the hair. Packets of powdered henna are commonly sold by grocers. There are two popular beliefs associated with henna. The first is more a social superstition. According to this belief, the unmarried girl who receives henna scrapings from a bride will herself be married soon (a similar superstition exists about a Christian bride's bouquet). The other belief about henna is that the darker the imprint left on the bride's palms, the more her mother-in-law will love her. This was probably devised to instill patience in the bride: to make her sit still for the henna to dry as well in dealing with her new family.

Sindoor

Vermilion, powdered red lead, applied as a dot on the forehead, like the regular 'bindi or in the parting of the hair by all suhagans. It is worn by married women, as a visible expression of their desire for their husbands' longevity. Traditionally therefore, widows did not wear vermilion.

Red is the colour of power. Vermilion is thus a symbol of the female energy of Parvati and Sati. Sati is believed to be the ideal wife, one who gave her life for her husbands' honour. Every Hindu wife is suppösed to emulate her. It is believed that Parvati protects all those men whose wives apply vermilion to their hair parting.

Sindoora is applied for the first time to a woman during the marriage ceremony when the bridegroom himself adorns her with it. Sindoora-dana, as this part of the marriage ceremony is called, is important today but is not mentioned before the Grihasutras indicating that it is a relatively new practice.

Ornaments

Gold ornaments are popular becausethe metal is believed to have the power purify anything it touches. Ornaments of gold and other metals, often combined with precious and semi-precious gems and beads, are popular with both men and women. Most ornaments are common across India, with variations in designs and material depending upon caste, religion, and geography. Gold is the most popular metal because it is believed to have the power to purify whatever it touches. For this reason, some gold ornament is usually worn against the skin at all times. Today ornaments are more popular with women than men, though, as paintings and sculptures attest men were once lavishly adorned too. In fact, the use of ornaments in India dates back to the ancient civilisations at Harrapa and Mohenjadaro.

Traditionally, ornaments had economic significance for women. The ornaments given to her at her wedding constituted a daughter's inheritance from her father. Customarily land and other property was divided among the sons, though this no longer holds true. In addition, a bride's ornaments were financial security throughout her life.

Nose Pin: More common than a nose ring, both are symbols of purity and marriage, though today many unmarried girls wear this adornment.

Necklace: These are popular across the country among girls and women of all ages. Necklaces are made of a variety of materials, ranging from glass beads to gold and diamonds. One special necklace, however, is the mangalasutra, worn only by married women. It is the Indian equivalent of the western wedding ring. Traditionally a woman wore it during her wedding ceremony and took it off only if her husband died. A lot of emotion and superstition focus on the mangalasutra.

Bangles: Worn on the wrist, bangles are believed to be protective bands and women always wore them as symbolic guards over their husbands. As with other ornaments, bangles today are worn by women of all ages all over India and are made of silver, gold, wood, glass, and plastic, among other materials.

Most married women wear glass bangles. However, in the eastern state of Bengal, they wear one shell bangle, called shakha, and one coral bangle, called paula, on each hand as a symbol of marriage. In the west, a bride from Punjab wears a set of ivory bangles on each hand for 21 days, six months, or a year after marriage, depending on family tradition. And a Rajasthani woman wears ivory bangles from her wrist to her upper arm for the rest of her life or till her husband is alive, although now this is rare.

Ear Rings: Rings, studs and other ornaments worn in the ears are popular all over the country. In fact, a girl's ears are usually pierced before her first birthday. Some ear ornaments are specifically worn by married women. For instance in the northern state of Kashmir, suhagans suspend gold earrings on long red threads from a hole pierced in the upper part of the ear. These earrings are called ateheru.

Other important ornaments are finger rings, toe rings and anklets. Rings for the fingers are again, of various materials and designs and worn by unmarried and married women. Since the ring has become a common adornment, it is no longer considered a symbol of marriage

However, toe rings and anklets are still worn mostly by married women. Ornaments for the feet are usually made of silver because gold, being a 'pure' metal, was not supposed to be worn on the feet. This privilege was given only to women of royal families.

In addition to these ornaments is the 'mangatika' or 'tikli'. This ornament, worn at the top of the forehead in the parting of the hair, is usually a small pendant on the end of a chain that is clasped to the hair. Although traditionally this ornament was also worn as a symbol of marriage, today it is not so commonly worn even by married women.

Nowadays, it is usually reserved for very special occasions. Traditionally, ornaments were considered a symbol of marriage and a widow did not adorn herself. Although this custom is still widely followed today, its observance is not strictly required.

Mangalasutra

Mangalasutra or thread of good will is a necklace worn specifically by married women as a symbol of their marriage.

Initially it was a thread that a bride and groom tied on each other signifying their union in good will. Now, the groom ties it around his bride's neck to indicate that he receives her from her father.

The most common Mangalasutra is made of two strings of small black beads with a pendant, usually of gold. The black beads are believed to act as protection against evil (see Nazar Utarna). Married women wear this to protect their marriage and the life of their husband. In southern India, the Mangalasutra is called 'tali'. It is a small gold ornament, strung on a cotton cord or a gold chain.

Traditionally, a married woman never removed her mangalasutra. If she became a widow, it was cut off. Nowadays however, this custom is not strictly applied, and not all married women wear a mangalasutra.

Puja

Puja is believed to be derived from the Dravidian word 'pu-chey', (flower action) or worship with the offering of flowers. Some trace it to the Dravidian word 'pusu', to anoint or smear with sandalwood paste or vermilion.

The term puja is now used to include all forms of ceremonial worship, ranging from the simple daily offerings of flowers, fruit, leaves, rice, sweetmeats and water to the deities in homes or temples, to the sacrifices of goats and chickens in temples dedicated to Kali, Durga and other female deities. This rite is performed, in its bloodless form, by all pious Hindus at least once a day.

There are three kinds of pujas: great, intermediate and small. A great puja is usually a community affair or performed during important occasions like religious festivals. This puja comprises of these steps:

Avahana-the invocation of the deity. Asana-a seat is offered to the deity.

Svagata-the deity is welcomed, asked Honour, worship, reverence.' to anoint or smear with sandalwood paste or vermilion. Puja about his journey and whether he faced any problem coming to the place of puja.

Padya-the feet of the deity are washed with water.

Arghya-a respectful offering of water is made to the god. This water is laced with sandalwood paste, vermilion and rice.

Achamania-water is then offered for washing the face and mouth of the deity.

Madhu-parka-a beverage made of honey, sugar, and milk is offered to the deity.

Snanajala-the deity is offered water for bathing. Bhushana abharanasya-clothes jewels and ornaments are offered next.

Gandha-sandalwood paste or any other fragrant object is offered. Akshata-grains of rice mixed with vermilion are offered.

Pushpanjali-flowers are offered.

Dhupa-incense is lit.

Dipa-the lamp is lit.

Naivedya-rice, fruit, butter and sugar are offered next.

Visarjana-the deity is finally bidden farewell.

At the end, arati is performed. An intermediate puja includes the steps from madhu-parka to naivedya and is performed during fasts or birthdays of deities. A small puja involves the steps from gandha to naivedya and is performed everyday. All pujas end with arati.

The object of performing the puja in this manner is to treat the deity as one would a guest, with honour and respect. In temples, the deities are treated as kings.

Though the steps of worship are the same for all deities, there is some difference in the puja of each. For instance, the kind of flowers offered is different for each deity.

Presently, a puja might also involve japa or meditation. A very important part of any puja is the applying of tilaka and the distribution of prasada to devotees.

A worshipper is required to be pure of body and mind. The Puranas lay more stress on the quality of devotion and good behaviour than on rigid puja procedures.

Puja originated as a substitute to homa and other Vedic sacrifices which could not be performed by women and Shudras and which required animal sacrifices.

Due to Dravidian, Buddhist and Jain influences which preached non-violence, the killing or sacrifice of animals was discontinued and with the development of iconography, idol worship and puja took the place of sacrifice. It was also recognized that worship was essential for all, whatever the gender or caste and therefore puja was formalised as a universal option instead of the exclusive homa.

Om

This mystic symbol is also the most powerful Hindu mantra. It is composed of five elements: the syllables a, u, m, the dot or bindu and sound (nada), and is revered as the representative of the gods. All aspects of speech are believed to have come forth from this word.

Om first appears in the Upanishads as a mystic monosyllable and is regarded as the seed or basis of all mantras. According to the Mandukya Upanishad, the past, present and the future are all included in this one sound and all that exists beyond these three forms of time is also implied in it. In fact this entire Upanishad deals with the understanding and symbolism of Om.

According to the Chandogya Upanishad, Prajapati created the first three Vedas, from which arose the three letters of 'aum'. This word is therefore a representation of the Supreme Being. Like the word Om, the Supreme Being is also indivisible.

Om is said to be the first sound to have come from the creator's mouth. Its sound is said to be the foundation of all sounds. It represents the whole phenomenon of sound producing and denotes all the possibilities of all the sounds that can be produced. 'A' is the root sound produced without touching any part of the tongue or palate, 'm' represents the last sound in the series, being pronounced by closed lips. Between these two sounds exist all others. The three constituents of Om also represent the deities of the Trimurti: 'a' represents Vishnu, 'u' Shiva and 'm' Brahma. Hence it also represents creation, preservation and destruction. The Vaishnavas and the Shaivas however describe it according to their own theological notions. The sound form of Om is identified as Vishnu by Vaishnavas and as Shiva by Shaivas.

The Shiva Purana claims that 'aum' arose from the five faces of Shiva. 'A' from the north, 'u' from the west, 'm' from the south, bindu from the east and sound from the middlemouth.

Because of this mystic importance, Om is uttered at the commencement of all prayers and religious ceremonies and at the beginning of all mantras. Om is also the mantra that yogis meditate with. Saying it aloud at the beginning or end of a lesson on Veda ensures that the student retains the knowledge imparted. Its utterance is said to counteract errors in the performance of sacrifices and protects the devotee against misfortune. Meditation on this sacred syllable is said to satisfy every need and leads to liberation.

This syllable was not allowed to be uttered by Shudras and women. The 'twice-born' (see Upanayanam) had to say it softly lest someone hear they considered unworthy. This is no longer the case and today, Om can be uttered by anyone. Om is now freely depicted on calendars, mandalas and even written at the head of letters, wedding invitations, doctoral. Theses and any important document. It has even found its way into popular film music and though no longer sacrosanct, is more potent than ever as the best known, most used symbol of Hinduism.

Arati

Arati Painful occurrence, injury, mischief, sickness. It is the most important ritual and is performed during almost all ceremonies and occasions. It involves the waving of an 'Arati plate' around a person or idol and is generally accompanied by the singing of songs in praise of that deity or person.

The arati plate is generally made of metal. On it must repose a lamp made of kneaded flour, mud or metal, filled with oil or ghee. A cotton wick is put into the oil and then lighted, or camphor is burnt instead. The plate also contains flowers, incense and akshata.

The purpose of performing arati is to ward off evil effects and the malefic influence of the 'evil eye' (see Nazar Utarna). Arati is hence performed on people of high social or economic status; small children during various ceremonies (see Sanskara); on people who are going on or are coming back from a long journey; on a bride and bridegroom when they enter their house for the first time; on grain (if one has had a good harvest); on animals or anything else of importance. It is also performed on newly acquired property, like a house or a tractor.

It is believed that the idol of a deity too is susceptible to the evil eye, and needs regular arati, with the singing of special arati songs. These songs laud the glory of the deities and describe the benefits that one might gain by praying to them.

Sometimes they also contain snippets of information on the life of the gods. Arati songs are particular to each deity. The most commonly sung arati is that to Vishnu. In most temples in India, arati is performed at least twice a day, after the ceremonial puja, which is the time when the largest number of devotees congregates.

While arati is being performed, the officiating priest waves the arati plate over the image of the deity. In doing so, the plate itself is said to acquire the radiance and the power of the deity.

The priest then takes the plate around to all those present as prasada. Arati The devotees cup their downturned hands over the flame and then raise their palms to their forehead. By doing this, it is believed that the purificatory blessing, passed from the deity's image to the flame, has now been passed to the devotee.

Ashram

Ashram or ashram is derived from cram meaning "to exert or labour". Ashram refers to a stage or phase in the life of a "twice-born" Hindu (see Upanayanam). There are four accepted ashrams.

These are Brahmacharya, for studying; Grihastha as a married man and householder; Vanaprastha, for performing penance in a forest; and Sanyasa, for renouncing worldly attachments.

According to the Manusmriti, the span of a human life is 100 years. Every Hindu man was to spend a part of his life in each of the four ashrams to fulfil his obligations: to the sages by studying, to his ancestors by raising male progeny, to the gods by performing penance and sacrifices, and for the salvation of his soul (see Moksha) by renouncing the world. This was assuming, of course that he lived his full life span.

In the Brahmacharya ashram, the adolescent is a celibate, religious student, called a brahmachari. After his. Ashram Upanayanam, a boy should live with and be devoted to his guru, and study the Vedas. He should live simply and humbly, follow the requisite customs, fast (see Vrata), and worship the gods. A student should study, practice penance, and tend the sacrificial fire.

This stage lasts for 12 years after the Upanayanam. However, if a student wishes to attain higher knowledge, he can continue in this stage until he is 31 years old. After completing his studies, the student gives guru dakshina and enters the next stage: the Grihastha ashram.

Grihastha means "householder". This stage begins with marriage (see Vivaha). The duties of a man in this phase of life include raising children, caring for the family, performing the five daily sacrifices and being a responsible member of society. He remains a Grihastha until his son's son is born, his hair has turned grey, or when he is about 50 years of age. Thereafter he enters the Vanaprastha.

As a Vanaprastha, a man lives in the forest, away from society, and prepares to renounce the world.

Vanaprastha means, "forest dweller". This stage indicates the beginning of old age. Living away from society in the forest, he should live simply, perform sacrifices, and prepare to renounce all worldly associations. These austerities are treated as penance for the sins committed, knowingly and unknowingly, as a householder. A man may take his wife to live with him in the Vanaprastha ashram, in which he stays until he is 70 years of age. He then returns from the forest and a purification ceremony initiates him into the Sanyasa ashram.

In this last phase, a man renounces wealth and worldly pleasures, and is called a sanyasi. He is completely detached from the rest of the world giving up even those family ties allowed in Vanaprastha. Instead, he seeks alms from householders and spends his time meditating, in an effort to achieve salvation (see Moksha).

Ashram phases followed a practical rationale. In the earlier part of his life, a man is more open to ideas and instruction, since his thoughts have not yet taken a definite shape. His energies and capabilities are unused and can be best-channeled into learning, as in the Brahmacharya ashram. The next 25 years are spent in the Grihastha ashram, where a man uses the knowledge he gained as a student to live a complete life and enjoys worldly pleasures without a sense of guilt.

After 50 years, the physical senses become weaker but inner strength increases. Satiated with the world, a man turns to the spiritual aspects of life in the Vanaprastha ashram. He goes on pilgrimages and slowly detaches himself from worldly affairs. After a point, he realizes that all attachments, even to his wife, are no longer necessary, and he has the ability to completely isolate himself mentally from the world. This marks the Sanyasa ashram, which is the last stage in a man's life.

The word brahmachari occurs in the Rig-Veda and Atharva Veda (see Veda), indicating that the system of ashrams dates back to the Vedic period.

There is controversy over which castes (see Varna) were supposed to follow the ashrams. According to some texts, the system was only for the Brahmins, while other sources say that any caste except the Shudras could practice it. Women, of any caste, were not permitted to follow this system. From childhood, they were taught how to keep a good home. The only ashram they observed was the Grihastha ashram.

There are three schools of thought about the observance of ashrams: Sammucchya or orderly co-ordination; Vikalpa or option; and Badha, annulment or contradiction. Sammucchya prescribes that a man should follow the four ashrams in order. Manu is the chief supporter of this view. The Vikalpa School preaches that a man can become a sanyasi immediately after the Brahmacharya Ashramashram or after the Grihastha ashram. But one can only become a sanyasi after studenthood if one has mastered control over one's senses and desires. The Vikalpa School does not consider Vanaprastha ashram a separate stage. This view is expounded in the Jabala Upanishad. The Badha school is held by the Gautama Dharmasutra and Baudhayana Dharmasutra (see Sutra). Followers of this school believe that there is just one ashram: the Grihastha ashram. They hold that the Brahmacharya ashram is only preparatory to the Grihastha ashram. Since each school is supported by Vedic text, all three are equally valid.

In the present day, few people find it practical to follow this system of life. Most people subscribe to the Badha school of thought, albeit unknowingly. First, they study, which prepares them for an occupation, and then marry. This stage, with slight variations according to different individuals, continues for the rest of their life.

Ganesha

Ganesha, also called Ganapati, is the god of wisdom, prudence, and salvation. Ga means "knowledge", na means "salvation", and isa and pati mean "lord". Ganesha is also said to mean 'lord of the ganas', Shiva's multitude of attendants. In south India, Ganesha is worshipped as Pillaiyar. Ganesha is represented as a short, pot-bellied man with an elephant's head with one tusk, four arms, and yellow skin. In mythology, the elephant symbolises devotion, patience and truth. His corpulent figure conveys prosperity as represented by the laddoos he is always shown with. In his hands, he holds a conch shell (see Shankha), a discus, a club, and a lotus. Ganesha rides on a mouse, signifying the unity of the small with the big.

He is also the most widely worshipped deity. All ceremonies, religious or secular, begin with an invocation to Ganesha. His image is printed on cards sent out for any happy occasion, like for the announcement of birth in the family or a wedding. Usually before beginning any religious writing, he is invoked with words 'Om Shri Ganeshaya Namaha', literally meaning "Ganesha, I pray to you". Most people, from student to shopkeeper, movie producer to nuclear scientist, begin work by invoking Ganesha, however briefly.

The Puranas differ considerably in their account of Ganesha's origin. He is not mentioned before 500 AD and was quite insignificant until the mid-15th century. According to the Padma Purana, he is the younger son of Shiva and Parvati.

According to the Shiva Purana, Ganesha was born of the scurf from Parvati's body. Once, when Shiva was away, Parvati wanted someone to guard the door while she bathed.

From the scurf of her body, she moulded a boy and gave him life. She told him to attend the door and not allow anyone to enter. Shiva returned, and when he tried to enter, was stopped by Ganesha. An argument arose, and enraged, Shiva cut off Ganesha's head and entered. When Parvati discovered that Shiva had entered by beheading her son, she was inconsolable. Repentant, Shiva ordered that the head of the first living being that was found should be brought to him. This happened to be the head of an elephant calf. He placed this on Ganesha's body and restored him to life.

A legend explains why Ganesha is worshipped before any other deity or prior to any important event. It happened that Shiva asked Kartikeya and Ganesha to circle the world and return. Kartikeya hurried off on his peacock, but Ganesha walked around Shiva and Parvati. He explained that for him, his parents constituted the world. Pleased, Shiva granted him a boon, saying that before undertaking any important task, people would pray to Ganesha. According to the Narasimha Purana, if he is not worshipped at the beginning of a ceremony, he creates obstacles for the performers. Therefore, no matter what the occasion or ceremony be, Ganesha is worshipped before all other deities. For this reason, he is called Vighneshwara, the remover of all obstacles. Ganesha's mouse, by gnawing its way through everything, is said to symbolise the god's ability to destroy all obstacles.

Ganesha is believed to have written the Mahabharata to Veda Vyasa's dictation. When the sage asked Ganesha to write down the epic, the learned god agreed on condition that his pen should not stop moving until the story was completed. Vyasa agreed but said that Ganesha should write only if he completely understood what was dictated. So whenever Ganesha stopped to consider Vyasa's complicated compositions, the sage would use the time to compose more verses. Ganesha is also known as Ekdanta, or the one with one tooth, because one of his tusks is broken. The reason for this, according to the Padma Purana, is that one-day when Shiva was sleeping, Parashurama came to visit him. However, Ganesha would not allow Parashurama in, for his father's sleep would be disturbed. When Parashurama insisted he be permitted entry, a fight broke out. In the course of their struggle, Parashurama threw his axe at Ganesha. This axe had been given to him by Shiva. Recognising the axe and out of reverence for his father, Ganesha refused to intercept the weapon. He bowed and took its impact on one of his tusks, which broke.

The Puranas say that the fourth day of the month, known as Chaturthi, is specially Ganesha auspicious for Ganesha worship. It is believed that Ganesha was born on the chaturthi of Bhadra. However every chaturthi is considered auspicious and Ganesha Chaturthi is an important festival in India, especially in Maharashtra.

Shiva Linga

Linga means "sign, mark, or symbol". It also refers to a digging implement or plough. Since the plough readies the earth for insemination, the word linga is also applied to the phallus, particularly as the phallic emblem of Shiva. Though Shiva is considered the destroyer (Trimurti) according to Hindu belief, creation (Srishti) follows destruction. The linga is therefore associated with the sexual union of Shiva and Parvati, which results in creation. It is a symbol of the procreative power of the universe.

Linga is derived from the Sanskrit word lingam, which comes from li meaning, "to dissolve" and gam "to move on". This refers to the belief that one appears as a being in the world and then dissolves back into the universe.

A Shivalinga is phallus-shaped and fixed on a base, which is shaped like a yoni. The structure symbolises the supreme creative energy. It is usually made of stone but can also be of wood, metal, crystal, and soapstone.

There are various explanations as to why the linga represents Shiva. According to the Vamana Purana, after Sati's death, Shiva went to mourn in the forest. Seeing the virile god, the wives of the sages living in the forest were enamoured and followed Shiva around. Enraged, the sages cursed Shiva, causing his phallus to fall off. The moment it touched the ground, his phallus started growing. Soon the earth, unable to support its weight, began quivering. When Brahma and Vishnu realised why the earth was shaking, they requested Shiva to take back his phallus. Shiva consented to do so on the condition that the sages worshipped his phallus. He pronounced that nothing would be impossible for one who worshipped the Shivalinga.

The Linga Purana says that Brahma and Vishnu were once arguing over who was the supreme being. Suddenly, there appeared Shivalinga before them a huge column of fire. Both of them decided to find one end each. Whoever returned first would be acknowledged as supreme. Vishnu assumed the form of a boar and dug into the earth. Brahma, in the form of a swan, flew upwards. They searched for days but in vain. Then Shiva appeared in the fiery column. Brahma and Vishnu realised their mistake and acknowledged Shiva as the Supreme Being. The Shivalinga represents that column of fire Jyotirlinga

The phallic cult appears to have been prevalent in India since the Indus Valley Civilisation (c. 3000 BC-700 BC). Emblems resembling the linga and yoni have been discovered in excavations of ancient cities. Although the Vedas are contemptuous of the linga because of its association with the phallus, later literature like the Puranas and tantric texts laud phallic worship. According to these texts, Brahma, Vishnu and other gods live at the site where Shiva exists in the form of a linga. The Vamana Purana says that during Chaturmasya, Shiva sleeps in all lingas. Therefore, praying at any linga during this time is considered especially auspicious. It is believed that the land within 100 cubits (about 160 ft.) of a linga is very sacred and called Shivasthala.

There are approximately 30 million lingas in temples and shrines all over India. Most of them are named after the place where they are located or after the person who established them there. Lingas fall into three categories. The first are the Anadi or Svayambhu, or self-generated lingas. They are believed to be the most sacred and are said to have fallen from heaven. Next are the Sithavara, or fixed lingas. These are crafted and then, after an initiation ceremony done according to Puranic rites, installed at a sacred location. It is believed that once a linga is installed, it should never be moved as that would result in a calamity. The third category of lingas are the Jangam, or movable lingas. These are the miniature lingas, which are made of stone, clay, metal, or other material. They are worn as amulets and necklaces, or worshipped in the homes of Shiva's devotees.

The actual worship of the Shivalinga begins with bathing it, first with water mixed with earth, and then with water mixed with the Panchagavya. The linga is then smeared with sandalwood paste. Next, flowers and leaves are offered especially bel leaves. Next comes the arati, followed by the devotee's prayer. Shiva is believed to be very hot-tempered: bathing the linga and applying sandalwood paste are believed to soothe him.

Jyotir Linga

The linga of light (see Shivalinga). There are believed to be 12 such lingas, the first to have appeared on earth, although their location is not consistently identified. Shiva is supposed to have first manifested himself as a jyotirlinga (lingodbhava) on the night of the Arudra nakshatra (see Panchanga). The jyotirlingas look like any other linga. However, it is believed that a person who has attained a certain degree of spirituality sees these lingas as columns of fire piercing through the earth. There are six definitely agreed-upon jyotirlingas: Kedar in the Himalayas; Bhimashankara at Dakini (Maharashtra); Tryambak on the banks of the River Gautami in Nasik (Maharashtra); Somnath at Saurashtra (Gujarat); Mahakala at Ujjain (Madhya Pradesh); Rameshvara at Jyotirlinga at Somnath.Setubandhana (Tamil Nadu). The other six commonly accepted jyotirlingas are Vishvanath at Varanasi (Uttar Pradesh); Amareshvara at Omkara; Vaidyanath at Parali (Maharashtra); Nagesh at Darukavan (Maharashtra); Ghushmesh at Shivalaya near Ellora (Maharashtra); Mallikarjuna at Srisailam (Andhra Pradesh).

Trimurti

Trimurti, meaning "having three forms", is the term applied to the three main Hindu gods: Brahma, Vishnu and Shiva. This Trimurti or triad represents all aspects of the Supreme Being.

The Trimurti is depicted as a single-bodied, three-headed man. This symbolises the fact that the three forms are aspects of one Supreme Being. Nothing in the universe is created, preserved, or destroyed without the mutual agreement and approval of the three aspects of the Supreme Being, for they are unitedly essential for the production and reproduction of all forms of life.

The concept of a divine triad is rooted in the earliest Hindu beliefs, which were largely fashioned by man's reaction to nature: reverence and fear, for there was no means of controlling it. As a result, he personified natural phenomena and worshipped them. The earliest triad comprised of Varuna, Mitra, and Aryaman, representing the basic aspects of nature like the day and night, earth and sky. As new concepts replaced older ones, this Triad gave way to another, of Vayu, Agni, and Surya. Later Indra replaced Vayu as a member of this highest order.

By Puranic times, the awesome aspects of nature had ceased to be so terrifying. Man was slowly understanding, and thereby learning to control, these phenomena. In addition, the earlier deities were limited: they could only control that aspect of nature, which they represented. A superior being was required who could control both nature and man, as well as provide solace. And so Brahma, Vishnu, and Shiva, (Creation, Preservation, and Destruction), came to be. According to the Puranas, the 'Trimurti' originated from Adishakti. However the widely accepted belief is that it represents earth, water, and fire. The earth has given rise to all living beings, and is therefore regarded as the divine creator. It is represented as Brahma, the creator. Brahma also represents the quality of passion or desire (rajo-guna), which is responsible for the creation of the world.

No living being can exist without water, which sustains life, and therefore preserves it. It is represented as Vishnu, the preserver. He also represents the quality of mercy (satva-guna), because of which the world is preserved.

Fire destroys life, and is responsible for the destruction of the world. It is represented as Shiva, the destroyer. He also represents the quality of darkness and wrath (tamo-guna), which is responsible for the annihilation of the universe.

The Trimurti is no longer worshipped as a deity. Even initially, it was worshipped as a representation of Brahman, the Supreme Being, and as a single entity. Later, each god was separated from the single entity and considered a deity in his own right. Today, Brahma is not worshipped, but Vishnu and Shiva both have large followings.

Brahma

Brahma is the god first represented in the Hindu Triad (see Trimurti). He is the creator of the universe and all living beings are said to have evolved from him.

Brahma is depicted as red in colour with four heads, bearded faces, and four arms. His hands hold a kamandalu, his bow or a rosary, a sacrificial ladle, and the Vedas. Brahma is sometimes depicted as sitting on a lotus. The four Vedas are said to have originated from his head. The four castes (see Varna) are also believed to have originated from Brahma: the Brahmins from his head, the Kshatriyas from his arms, the Vaishyas from his thighs, and the Shudras from his feet.

Brahma was assigned the work of creating the universe by Brahman, the Supreme Being to assist him in this endeavour, Brahma created 10 Prajapatis, who are believed to be the fathers of the human race. Their names, according to the Manusmriti are Marichi, Atri, Angirasa, Pulastya, Pulaha, Kratu, Vasishtha, Prachetas or Daksha, Bhrigu, and Narada. These Prajapatis are believed to be born from the mind of Brahma. He is also believed to have created the Saptarishi, or seven great sages, to help him create the universe.

The world is said to exist for one kalpa, or one day in the life of Brahma. At the end of a kalpa, the whole world is destroyed. Brahma is then believed to go to sleep for one night, which is as long as a kalpa. When he awakens, he recreates the world. This process (pralaya) is repeated for 100 years of Brahma, which is the life span of one Brahma. Then everything dissolves into its constituent elements. This whole process is repeated unendingly.

Brahma's abode is known as Brahmaloka, which is said to contain all the splendors of earth and of the heavens of the other gods. His vehicle or vahana is a white swan or goose, which has magical abilities: it can separate soma and milk from water. Therefore this bird is also a symbol of sifting good from evil. Unlike other deities, Brahma does not have a weapon. Being the lord of sacrifices, he is identified with a sceptre in the form of a spoon. However, some sources say that his weapon is the bow, which he is sometimes shown with.

His consort is Saraswati, the goddess of wisdom and science, the mother of the Vedas, and the inventor of the Devanagari script. She is represented as a fair woman with four arms, dressed in white and seated on a white lotus. As goddess of the arts, she is shown playing, or holding, a veena. In one of her right hands, she holds a book of palm leaves, and in the other, a lotus. In her left hands, she has a string of pearls, and a damaru.

Although Brahma is considered equal to Vishnu and Shiva, he is currently not widely worshipped. There are only two temples in India dedicated to him, one at Pushkar near Ajmer in Rajasthan, and the other at Khedabrahma in Kerala. However, he is only worshipped by sanyasi (see Ashram). Apart from these temples, there are also some verses describing which are chanted by Brahmins every morning.

There are several reasons that Brahma is no longer worshipped. One belief is that being the creator, his work is complete, at least for the time being. Therefore he is not as revered as the other two gods of the triad (see Trimurti).

Another reason is based on a legend in the Skanda Purana. Attempting to prove his superiority, Brahma lied to Vishnu, while the ketaki flower stood false witness for him. For this, Brahma was cursed by Shiva that he would never be worshipped on earth, nor would the ketaki ever be offered in worship (see Jyotirlinga). And so it is, since time immemorial. Some of Vishnu's incarnations or avatars, like Matsya, Kurma, and Varaha, were earlier ascribed to Brahma, for both Vishnu and Brahma were described as 'Narayana', one who moves on the waters. However, as Brahma's importance declined and Vishnu's rose, these avataras were transferred to Vishnu.

According to Manu and the Shatapatha Brahmana (see Brahmana), Brahma emerged from Brahman, the Supreme Being. Brahman wanted to create the universe. To this end, he first created water, into which he deposited his seed. This seed became a golden egg, and from this egg, he was born as Brahma or Hiranyagarbha, born of the golden egg. Another account of Brahma's origin is found in the Mahabharata. It says that Brahma arose from a 1,000-petalled lotus that grew out of Vishnu's navel. According to the Puranas, Brahma is said to be the son of the Supreme Being and his female energy, Maya. Other sources however, state that Maya is the wife or daughter of Brahma.

Originally, Brahma is said to have five heads. The Matsya Purana explains the reason for this. It states that Brahma created a woman, known by different names: Satarupa, Saraswati, Sandhya or Brahmi. Because of her beauty, he fell in love with her and stared at her longingly. To avoid his gaze, she moved to his left, then behind him, and then to his right. But a head sprang up wherever she moved to enable Brahma to continue looking at her. In desperation, she jumped into the air, but a fifth head appeared on top. Brahma then asked her to help him create the universe. He lived with her for 100 divine years at the end of which Manu was born.

There are different explanations for why Brahma only has four heads now, as opposed to his original five. According to the Puranas, Brahma and Vishnu were once arguing over who was superior of the two. They discovered from the Vedas that Shiva was the Supreme Being. Brahma however, spoke disparagingly about Shiva. In anger, Shiva cut off the head, which had spoken, and therefore Brahma was left with four heads.

Vishnu

Vishnu, (from the Sanskrit root 'vish', to pervade) is the second god of the Hindu triad (see Trimurti) and as the preserver of the universe represents mercy and goodness. He is said to be the cosmic ocean nara, meaning water, which was the only thing in existence before the creation of the universe. For this reason, he is also known as Narayana, or "one who moves on the waters".

Vishnu is represented as a dark man with four arms. In one hand, he holds a club, in another a conch shell (see Shankha), in the third, a discus (Chakra), and in the fourth, a lotus (Kamal). He is usually dressed in yellow robes and therefore also known as Pitambara, or "one with yellow garments". The river Ganga is said to originate from Vishnu's feet. He is also depicted as resting on a coiled serpent Seshnaga or Ananta (see Snake Worship) which floats on the cosmic ocean. According to the Mahabharata, Vishnu's abode, Vaikuntha, is made of gold and jewels. However, he is usually depicted reclining on Seshnaga, who floats on the cosmic ocean. His vehicle is Garuda, and his weapon is the discus, Sudarshana Chakra.

Vishnu's consort is Lakshmi, the goddess of wealth and fortune. She is believed to have emerged from the samudra manthan, and considered to be the daughter of Bhrigu and Khyati. She is a fair woman who wears red clothes, and is always depicted as sitting or standing on a red lotus (see Kamal). She is often worshipped by herself as Sri, Bhagya Lakshmi and Dhana Lakshmi, the goddess of all things auspicious, of good fortune and wealth.

Vishnu is a very popular deity and is widely worshipped. The Vishnu Purana details the benefits of his worship. He has a 1,000 names, whose repetition is believed to bring great merit. He is also worshipped as Satyanarayana. Apart from this, there are many other vratas and festivals related to Vishnu. A variety of ammonite fossils, the shalgrama shila, is believed to be a form of Vishnu and is widely worshipped. There is also a specific arati associated with Vishnu, which eulogises him. Those who believe in the supremacy of Vishnu over Shiva are known as Vaishnava. They can be distinguished by their tilaka, which is different from that of Shaiva. Vishnu as the preserver is not mentioned in the Rig Veda (see Veda). In the text, he is considered to be the manifestation of solar energy. His role as the preserver was established by the Puranas.

The Padma Purana describes him as the Supreme Being, identifying him with Brahman. It states that Vishnu, wanted to create the universe. To this end, he divided himself into three parts: from his rightside, he produced Brahma the creator; from his left, he produced Vishnu, the preserver; and from his middle, he produced Shiva, the destroyer.

Vishnu has several avataras. According to the Puranas and the Mahabharata, whenever there is an imbalance between good and evil on the earth, Vishnu, as the preserver, is born to re-establish the balance. The Bhagavata Purana, among other sources, say that there are 22 avataras of Vishnu. However, the widely accepted belief is that there are 10 avataras, known as the Dashavatara. All the 10 avataras are revered, but the seventh, Rama (see Ramayana), and eighth, Krishna, are believed to be great heroes and worshipped as gods in their own right. Of Vishnu's 10 avataras, the first four are believed to have occurred in the Krita Yuga, the fifth, sixth, and seventh in the Treta Yuga, the eighth in the Dvapara Yuga, and the ninth in the Kali Yuga. The tenth avatara has yet to appear, and it is believed that he will come at the end of the Kali Yuga. The avataras are Matsya, the fish, which appeared in the Krita Yuga. Vishnu assumed this form to save Vaivasvata, the seventh Manu, from the universal deluge that occurs after each Manvantara.

Kurma, the tortoise, assumed to act as a pivot for Mount Mandara and protect the earth during the samudra manthan.

Varaha, the boar. Hiranyaksha, a demon, had dragged the earth to the bottom of the sea. He had also obtained a boon from Brahma, ensuring that he would not be killed by any of the creatures he named. However, while doing so, Hiranyaksha omitted to name the boar. Therefore Vishnu assumed this form and dived to the depths of the ocean. After defeating the demon, he balanced the earth on the horn above his snout and surfaced safely with her. Narasimha, the man-lion. After severe penance, another demon, Hiranyakashipu, had been granted a boon by Brahma, Vishnu according to which he would die neither by day nor at night, neither indoors or outdoors, and be killed by neither man nor beast. Considering himself invincible, he became a tyrant. In order to kill him, Vishnu assumed the form of a man-lion. He killed Hirañyakashipu at twilight on the threshold of his palace. But the demon's son Prahalada was an ardent devotee of Vishnu and won his favour. Vamana, the dwarf, was the first avatara to appear in the Treta Yuga. Vishnu assumed this fifth incarnation to subdue Mahabali, a demon king, who by performing severe austerities, controlled heaven, earth and hell. The gods were shorn of their power and significance. When Mahabali performed an important sacrifice, Vishnu turned up as Vamana, the dwarf-Brahmin son of Kashyapa and Aditi. He asked Bali to grant him as much land as he could cover in three steps. When Bali agreed, as he was bound to by custom, Vamana grew to a gigantic size, and in two.

Vishnu steps covered heaven and the earth. However, he left Patala, the netherworld for Bali, who had to offer his own head for Vishnu's third step to be completed onam. Parashurama, 'Rama of the axe', a Brahmin, born to the sage Jamadagni and his wife Renuka. His father suspected his mother of infidelity and ordered Parashurama to behead her with his axe. He did so and when his father granted him a boon in reward, he asked to have his mother back. She was deified later as an incarnation of Parvati. Parashurama was born to annihilate the Kshatriyas, who had fallen into evil and oppressive ways, led by the insolent king of Mahishmati, Kiratarjuna, who eventually killed Jamadagni. In revenge, Parashurama tried to wipe out the entire Kshatriya caste (see Varna) 21 times, but each time a few escaped. According to the Puranas, Parashurama was successful in his last attempt (see Kshatriya). He then went to the shores of the Arabian Sea and flung his bloodied axe into the ocean, which yielded up the verdant, spice-laden land of Kerala. Thereafter, he retired to meditate. He appears briefly both in the Ramayana and the Mahabharata.

Rama, the ascetic-prince of Ayodhya, who is widely worshipped in the north as the embodiment of human perfection and the killer of the demon-king Ravana, but remains controversial for the treatment of his consort, Sita. After being kidnapped by Ravana and rescued by Rama, she had to undergo a public ordeal by fire to prove that she had been left chaste while in capture. Though Agni himself returned her with all honour to Rama, he nevertheless banished her to the forest some years later, while she was pregnant, because of the ramblings of a drunk washerman berating his own wife (see Ramayana). Krishna, the eighth incarnation. Krishna was born to destroy his evil uncle, Kansha, who had usurped the throne of Mathura. He played a crucial role in the Mahabharata and fully revealed his divinity as the expounder of the Bhagavad Gita to Arjuna. Like Rama, Krishna is a popular deity and is widely worshipped. His death is believed to have marked the beginning of the Kali Yuga.

Buddha, the enlightened one, founder of Buddhism. The inclusion of the Buddha as one of the Dashavataras is clearly an attempt to incorporate Buddhism into Hinduism. Vishnu is believed to have taken the form of the Buddha to encourage evil people to despise the Vedas, and reject the caste system and the deities. In this way they would effect their own destruction. This incarnation appeared in the Kali Yuga.

Kalki, the rider. This tenth incarnation of Vishnu is yet to appear on earth. It is believed that Kalki will ride a white horse and wield a flaming sword with which he will destroy all evil. The coming of this fierce horseman will mark the end of this present age of evil, the Kali Yuga, after which purity will reign once again in another Treta Yuga.

Mahesh

Shiva literally means "auspiciousness, welfare". He is the third god of the Hindu Triad (see Trimurti) and is the destroyer of all evil. He represents darkness (tamas), and is said to be the 'angry god'. However, according to Hinduism, creation follows destruction. Therefore Shiva is also regarded as a reproductive power, which restores what has been dissolved. As one who restores, he is represented as the linga or phallus (see Shivalinga), a symbol of regeneration.

He has a 1,008 names, including Mahadeva (the great god), Mahesh, Rudra, Neelkantha (the blue-throated one), and Ishwar (the supreme god). He is also called Mahayogi, or the great ascetic, who symbolises the highest form of austere penance and abstract meditation, which results in salvation (see Moksha).

Shiva is believed to exist in many forms. His most common depiction is as a dark-skinned ascetic with a blue throat. Usually seated cross-legged on a tiger skin, Shiva's hair is matted and coiled on his head, adorned with a snake and a crescent moon. Ganga is always depicted flowing out of his topknot. Shiva has four arms and three eyes. The third eye, in the middle of his forehead, is always closed and only opens to annihilate an evil doer. A garland of skulls, rudraksha beads, or a snake hang from his neck. Shiva also wears snakes as armlets and bracelets. The serpent race, despised and feared by all other creatures, found a place of honour on Shiva's sacred person, simply because he was moved by their plight.

In one hand, Shiva holds his trishul, the Pinaka. The trishul usually has a damaru or waisted drum tied to it. In another hand, he holds a conch shell (see Shankha), and in the third, a rudraksha rosary, a club, or a bow. One hand is usually empty, raised in a gesture of blessing and protection. The other points to his feet, where the devotee is assured of salvation. He wears a tiger or leopard skin around his waist, and his upper body is usually bare, but smeared with ashes, as befits an ascetic. His third eye is believed to have appeared when Parvati, in a playful mood, covered his eyes with her hands. Immediately, the universe was plunged into darkness and there was chaos. To restore order, Shiva formed another eye on his forehead, from which emerged fire to restore light. The light from this eye is believed to be very powerful, and therefore destructive. Shiva opens his third eye only in anger, and the offender is burnt to cinders.

The name Shiva does not appear in the Vedas. However he is identified with the Vedic god Rudra, lord of songs, sacrifices, nourishment, the healer of diseases and provider of property. According to the Shiva Purana, Shiva is said to have five faces, corresponding to his five tasks, the panchakriya: creation, establishment, destruction, oblivion, and grace. His five faces are associated with the creation of the sacred syllable Om.

Shiva is said to live on Mount Kailash, a mountain in the Himalayas. His vehicle is Nandi the bull and his weapon, the trishul. Shiva's consort is Parvati, who is also believed to be a part of Shiva. One of the most popular forms of Shiva is that of Ardhanarishvara.

According to a story in the Puranas, Brahma was unsuccessful at creation. He propitiated Shiva who took this form and separated Parvati from his body. Parvati has many incarnations, like Kali, Durga, and Uma. Their sons are Kartikeya and Ganesha.

Shiva is believed to have a large number of attendants, called ganas. These mythological beings have human bodies with animal heads. Shiva's son Ganesha is the leader of the ganas.

Across the country, there are hundreds of temples and shrines dedicated to Shiva. He is usually worshipped in the form of a shivalinga, and also as an idol. He is worshipped by offering flowers, excepting the Ketaki Brahma Bel leaves, milk, and sandalwood paste are also pleasing to him. There is a special arati to Shiva and many hymns and poems in his praise.

There are many stories in the Puranas about the origin of Shiva. According to the Vishnu Purana, at the beginning of this kalpa Brahma wanted a child and meditated for one. Presently, a child appeared on his lap and started crying. When asked by Brahma why he was crying, the child replied that it was because he did not have a name. Brahma then named him Rudra, meaning "howler". However the child cried seven more times and was given seven more names.

Shiva therefore has eight forms: Rudra, Sharva, Bhava, Ugra, Bhima, Pasupathy, Ishana, and Mahadeva, which, according to the Shiva Purana, correspond to the earth, water, fire, wind, sky, a yogi called Kshetragya, the sun, and the moon respectively. During the samudra manthan, when poison was churned out of the ocean, Shiva is said to have swallowed it to save the world from destruction. As he drank the poison, Parvati clasped his throat tightly so that the poison remained there and darkened his neck. Because of this, he is known as

Neelkantha, the blue-necked one. Shiva granted a peculiar boon to a demon called Bhasmasura, who wished that if he placed his hand on anyone's head, that person would turn to ashes.

No sooner was his wish granted than he menaced Shiva himself, who took to his heels and was saved by Vishnu, in the form of Mohini, the enchantress Induced by Mohini to join her dance, the demon soon killed himself by placing his hand on his own head. Shiva and Mohini combined their energies and produced Hari-hara putra (Vishnu-Shiva's son), later identified with Sasta or Aiyappa, a celibate tribal deity in Kerala, whose cult is now enormous in India.. He lives on in Mohiniattam, the feminine counterpart of Kerala's dance theatre, Kathakali.

Shiva is the creator of dance and of the first 16 rhythmic syllables ever uttered, from which the Sanskrit language was born. His dance of anger is called the Roudra Tandava and his dance of joy, the Ananda Tandava. All the gods and sages were present when he first danced the Nadanta Tandava, a characteristically vigorous dance, and they begged him to dance again.

Shiva promised to do so in the hearts of his devotees and in a sacred grove in Tamil Nadu, where the great temple of Chidambaram was built, the only one in all India dedicated to Shiva as Nataraja, the lord of dance. It is believed that on the 13th day of each bright lunar fortnight (see Hindu Calendar), after 6 o'clock in the evening, falls a sacred hour called Pradosha. Worshipping Shiva at this time is akin to worshipping all the powers. Shiva the universe, for this is the time when all the gods are believed to have assembled on Kailash to lose them in the ecstasy of Nataraja's dance.

Superstitions

Like people the world over, most Indians are highly superstitious. Many aspects of their life is linked to some superstitious belief and some of these are even mentioned in religious literature. Most superstitions were born with an aim to protect and were based on sound reasoning. With time, however, much of the reasoning has been forgotten and the beliefs appear unsubstantiated. But there are also many beliefs that have no reason behind them.Of superstitions, those believed to portend future happenings are considered important. Anything from the call of a bird to the falling of utensils is considered an omen.

Many omens relate to animals, birds and reptiles. The most auspicious omen is to see an elephant when one is on a journey, for it represents Ganesha, the god of good luck and the remover of obstacles. The cawing of a crow in one's house foretells the coming of guests. A peacock seen while on a journey is said to be good but hearing its shrill call indicates robbery by highway men. A dog howling near a sick man's chamber predicts his death. Seeing a cat or a cow's face early in the morning brings ill luck. A sparrow is encouraged to build a nest in a new house for good fortune. A yellow or white substance dropped on the threshold by a cow indicates the gain of gold or silver. To protect cats, which were useful in destroying rats and other pests, the Brahmins imposed a ban on its killing, to be redeemed expensively with an offering of a golden cat to a Brahmin.

The wall lizard boasts the most superstitions. Every movement of this reptile holds some significance, indeed, a science called the Gowli Shastra enumerates these. The colour, spots, stripes, chirping or twittering of the lizard and where it falls on a person's body are said to indicate future happenings.

Leaving on a journey is a very important occasion and precautions have to be taken depending on the importance of the journey. The astrologer is consulted to select an auspicious time to begin. There are several beliefs linked with starting on a journey. A sneeze or terms of dissuasion like "Stay, don't go", or "Where are you going?" used accidentally by someone at this crucial time are ill omens and a person must start again if this happens. It is considered auspicious to see cowdung, cereals, paddy, cotton, hay or a newly wed before starting on a journey.

There is also a strong belief in the power of dreams, as divine warnings. Dreaming of gods, demons, auspicious animals or any other auspicious thing is good. Whereas, dreaming of gold or iron, falling stars or earthquakes is bad. There are many people who claim to be able to decipher dreams and their talents are employed by believers.

Daily life is governed by do's and don'ts for each day of the week. Thus Monday is not an auspicious day for shaving or Tuesday for washing one's hair or Saturday for buying oil.

Most of these superstitions are linked to the respective planets presiding over each day (see Hindu Calendar). The kitchen is virtually the temple of a Hindu home and the highest levels of hygiene are expected to be maintained. Most Hindus do not enter a kitchen with their shoes on, as it is said to induce the wrath of the gods. This belief originated when kitchens were traditional, with mud stoves in a corner. Families sat on the floor and ate off plates or leaves placed on the ground. This required the kitchen to be absolutely clean. Another belief linked to this was that before eating their food, people sprinkled water around their plates supposedly in remembrance of their ancestors. Actually, the water around the plate acted as a barrier against the ants and insects that crept about the kitchen floor.

There are other common beliefs devised for protection. One should neither clip one's nails at night nor use scissors for fear of angering evil spirits. This diktat prevailed in the days before electricity as a deterrent to injuring one's fingers in the dark, but many mothers still enforce this rule at home, the reasoning being lost in time. A persistent belief is that one must not sweep the floor after the sunsets, otherwise Lakshmi would also be swept away. This too is a pre-electricity precaution against inadvertently losing fallen valuables.

Some other popular superstitions are if a person hiccups, it is believed that someone is talking about him or remembering him. One theory holds that this is a way of keeping alive the memory of family members who are travelling or live far away.

The falling of utensils on the floor foretells the coming of guests. Breaking a mirror is extremely inauspicious. This belief is meant to urge caution while handling something as fragile and potentially injurious as glass, as most people were barefoot inside the house.

To protect foolhardy wanderers from snakes and other lurking dangers, it is held that evil spirits infest trees and empty houses, resulting in a lively fear of going to these places at night.

There are also many superstitions related to a pregnant woman, which originated for her protection. Some of the most widespread are: A pregnant woman should not wander out alone at night or enter an empty house because evil spirits might infest her. The husband of a pregnant woman should not build a house as that might rouse the wrath of the earth spirit. He must not bathe in the sea for the seashore is thick with ghosts. All these rules were imposed on people, as ways to protect the prospective mother and father against any dangers. In addition, when a woman is pregnant, her husband's presence is important for moral support and help with the chores. Therefore unnecessary journeys or the involvement of the husband in any major tasks is postponed.

There is a superstitious belief in some parts of the country that when a child is born, its father must not see its face till it is six months old as it might lead to ill-luck. Therefore some time before the child is born, the mother is sent to her parents' home. This serves two practical purposes: the mother is pampered by her own mother, and so is the baby, as all the mother's attention is centred on her child, with no distractions of looking after her husband or doing other chores.

In the post-partum period and while menstruating, a woman is debarred from attending any ceremony. She is also not allowed to enter the kitchen or mingle with others, as she is believed to be impure and unclean at this time. The underlying idea is that a woman at this time requires complete rest and a religious edict protects her from unnecessary exertion. Personal appearance, gait, manner and speech are all said to indicate the future. The location, size and number of moles on the body brand someone as auspicious or inauspicious. It is believed that the presence of a sign resembling a wheel on any part of the body is extremely lucky. The right side of the body is considered to be more auspicious than the left. This is so because the left hand is used for all 'unclean' actions. Therefore a man ensures that he first puts his right foot into a house. All ceremonies are conducted with the right hand. Gifts are also given and accepted only with the right hand. When the right hand itches it means good luck and the reverse applies to the left hand. This relates simply to the clean (right) and unclean (left) sides of the body. Paradoxically, twitching of the right eye is considered inauspicious and of e left eye, auspicious.

Marriage, one of the most important events in life, has many superstitions associated with it. It is believed that a new bride brings luck and prosperity with her and that her fortune is mingled with that of the groom's family. For the first one year of marriage, all happenings in the family are monitored and accordingly the bride is termed auspicious or inauspicious. Any good or bad occurrence in the family during this time is said to result from the presence of the bride. A new bride is however, generally considered to be auspicious.

On the other hand, widows were considered especially inauspicious. This belief probably stemmed from her traditional financial powerlessness. Widows were debarred from attending any auspicious ceremony, as it was believed that their presence might negatively affect its success. A new bride was protected from the glance of a widow for the same reason. Therefore widows had separate areas to live in where they cooked their own food. They were not allowed to eat meat, onions, garlic or spices for these were considered excitatory, and a widow was supposed to suppress, not inflame her passions. Most of these rules have fallen into disuse now.

A barren woman is still considered inauspicious by many, given the patriarchal social system and its emphasis on children especially sons. Such a woman, though not debarred, is avoided during all auspicious occasions by those with a rigidly traditional outlook. During the nuptial bath and other ceremonies related to marriage, which require the presence of suhagans, it is generally women with sons, pregnant women or the newly married, who participate. A barren woman is not brought near a bride or child for whom a ceremony is being performed for fear that her ill fate will rub off onto them.

People are also very superstitious about their mornings. It is believed that the whole day is affected by the morning. Seeing a bride, the statue of a deity, or a Brahmin is considered auspicious, while sighting widows, barren women and unlucky people is considered inauspicious. To avoid the risk of seeing anything unpleasant in the morning, many people look at the palms of their hands as soon as they awaken and recite this invocation: "In the tips of the fingers resides Lakshmi; in the middle, Saraswati; in the palm of the hand resides Parvati; looking at my hands, I begin my day."

There are also many superstitions related to death. The most important among them is that food is not cooked in a house where death has occurred. At this time, all the members of the family are considered to be unclean and do not enter the kitchen. Neighbours are expected to cook and send food for them. This belief probably results from the fact that when a death occurs in a family, the relatives of the deceased are, and have neither the desire nor the time to cook. This practice also reinforces the sense of community at times of trouble.

Tilak

Tilaka or Pundraka is a ritual mark on the forehead, of sandalwood paste, coloured earth or vermilion. It is a sign of sectarian distinction, and of auspiciousness.

The spot between the brows where the tilaka is applied is considered the seat of latent wisdom and mental concentration, which is very important for worship. This is the spot on which yogis meditate to become one with Brahman. It also indicates the point at which the spiritual eye opens. All thoughts and actions are said to be governed by this spot. The tilaka, initially of either sandalwood paste or some other cool substance, was applied on this spot to keep it cool so that man could be of stable mind and thought. Most Hindu scholars, even today, sport a sandalwood paste tilaka.

As various sects arose with time, it was the tilaka that indicated sectarian distinctions. For instance, the devotees of Shiva apply three horizontal lines of vibhuti (sacred ash) on their forehead. This is to remind themselves of the God's threefold nature of creation, preservation and destruction. The ashes also indicate his main function: destruction.

The devotees of Vishnu apply three vertical lines of sandalwood. This is the sign of his preservation and protection. The devotees of Devi (Shakti) apply the kumkum, which is a round or slightly elongated red mark. This evokes the supreme power of Adi Parashakti, the Universal Mother, from whom all life and energy, even the power of the Trinity (see Trimurti), is believed to emanate.

For the lay worshipper, the most common tilaka these days is the one applied ritually after a puja or arati, of red kumkum or sandalwood paste.

Shika

A tuft or lock of hair on the crown of the head, a crest, topknot, plume.

From the time of the Vedas, the shikha was a distinguishing feature of the Brahmins, Kshatriyas and Vaishyas. It signified the 'twice-born' or all those Upanayanam has been performed. At the time of Chudakarana, a tuft of hair was left on the head, never to be cut. This shikha covered a large part of the brain. According to Sushruta, the reason that a few tufts are left on the head is that at the crown, an artery joins a critical nerve juncture. Since an injury to this part of the head is believed to be fatal, it was considered necessary to protect the area by keeping a tuft of hair over it. The shikha was a symbol of superiority and of cleanliness.

Any religious or auspicious ceremony required the shikha to be tied in a knot. The knot was tied to the accompaniment of the Gayatri Mantra. An untied shikha was a symbol of disgrace, impurity and mourning.

Tying of the Shikha

During funeral and death ceremonies (see Antyeshti) it was left untied. Presently only Brahmins wear the shikha, especially by practicing priests. The shikha on the top of the head represents the shikhara that is raised atop all Hindu temples. Temple architecture is based on the vastu purusha mandala or Mystic Square, a diagram said to represent the Supreme Being (see Brahman). A temple is considered a symbolic depiction of the Supreme Being, whose head is represented as the shikhara of the temple. And in men, the shikha symbolises the brahamanda on which Brahma is said to reside.

Swastika

Swastika is derived from su (well), asti (is) and ka (a noun ending). It means 'it is well' and hence signifies happiness, pleasure and good luck. It is a mystic symbol or figure, which is sacred to most Indian sects. The symbol is shaped like a Greek cross with the extremities of the four arms bent round in the same direction. A swastika is of two types: the right-handed or male, representing the vernal sun and the god Ganesha. In this Svastika, the extremities of the arms of the cross bend clockwise. It is considered auspicious. The left-handed or female cross represents the autumnal sun and the goddess Kali. Its arms are bent in an anti-clockwise direction and it is believed to be inauspicious.

The auspicious symbol is used by the religiously inclined on the opening page of account books. It is also used in ceremonies related to marriage, tonsure, the worship of Svastika Lakshmi and during the invocation of the nine planets (see Janmapatri). It is worshipped as a symbol of the sun, Ganesha and the serpent kingdom (see Snake Worship). According to the Vayu Purana, serpents bear the sign of a swastika on their hoods. In festivals and on auspicious occasions, the swastika is drawn on the floor of the house. A yogasana (see Yoga) is named after this symbol, and is considered to signify or evoke good luck. Many theories have been propounded about the swastika.

It has been described as the emblem of Zeus, the god of sun, fire and rain, the Trinity (see Trimurti) and also the Supreme Being, Brahman. Most scholars however regard it as a fire or solar symbol and hence it is called the solar or fire cross. It is believed to be derived from the chakra, which is symbolically reduced to four spokes and set at right angles. It also represents the fire-making apparatus of old times, used for kindling the sacrificial fire (homagni). In the Vedas it is referred to as 'the wheel of the sun'. It indicates cosmic procession and evolution around a fixed centre. It also represents the principle of life and movement that transfers the dwarf to the giant and the microcosm to the macrocosm, named rkvam, the rhythmic movement.

This symbol has been in use since the time of the Indus Valley Civilisation (c. 3000 BC-700 BC). A large number of seals discovered in Harappa, one of its prime sites, bear swastika designs. Since then, it has represented good luck. Panini, the great Indian scholar, used this symbol as the name of a sign in his grammatical treatise, Ashtadhyayi ('Eight Chapters', 450 BC). The symbol is not exclusive to India and is known all over the world. Proof of its widespread use has been found in archaeological discoveries in Egypt, China, Greece and Mexico. Curiously, the inauspicious anti-clockwise swastika was used by Hitler as the Nazi emblem.

Moksha

Moksha, release from worldly existence, final or eternal emancipation, the liberation of the soul from the recurring cycle of birth and death, salvation. It is the spiritual union of the soul, atman, with the supreme soul, Brahman. The Hindus believe in transmigration of the soul: that the soul does not die with the body, but, after a while, takes birth again in some other form. Immediately after death, the soul is believed to travel to Yamapuri, the court of Yama. Here Chitragupta, his record keeper, reads out an account of all the good and the bad deeds committed by that soul during its life.

The soul is then given a chance to explain itself. After this, Yama passes judgement and the soul is accordingly sent to heaven or hell. The Hindu hell, however, is not an endless ordeal of misery, as in other beliefs. After serving its allotted term of punishment, the soul is taken to heaven on account of its good deeds. Heaven is also not an endless term of bliss. Hence, after its term there, the soul is reborn in this world in one form or another and the whole process is repeated.

The form in which the soul is reborn depends on the deeds of its previous life. If a soul has done well, it may be born a king. If evil and cruel, it could be reborn as a jackal or a donkey. A soul is not always reborn on earth. It may reappear in any of the fourteen worlds, as a god, a demon or a snake

Birth in any form is considered undesirable, as it brings pain and suffering. Therefore everyone is exhorted to achieve liberation from this tiresome cycle of birth and death, by attaining moksha.

Such a state is acquired only by a very saintly person, who has become a perfected soul with no worldly requirements and true knowledge of Brahman.

All sacrifices, pilgrimages, and acts of charity that one performs are directed towards improving the self in order to be born in a higher form, which ultimately leads to the final destination, moksha. This is also the reason why penance or prayashchit is considered so important.

The three paths to moksha are:

1. Bhakti yoga (through devotion), the most popular, in which a devotee chooses a 'personal' deity and prays to it with intense love and devotion.
2. Karma yoga (through action), by devoting one's life to good causes, to the betterment of society with sincerity and faith.
3. Gyana yoga (through knowledge), through perfecting oneself in a particular discipline and meditating on it. Their belief in the transmigration of souls makes the Hindus a fatalistic people. As a result, all one's troubles are attributed to the bad deeds of one's previous birth.

Varna

Literally "colour", the caste system. The word "caste" itself comes from the Portuguese and Spanish casta meaning "lineage, race or breed". In Sanskrit, the equivalent word is jati meaning "birth".

This system was originally introduced to India by the Aryans who arrived with three loose social divisions: priests, warriors and commoners. This system was a division of labour and not a social hierarchy. Caste was dictated by occupation and intercaste mobility was the norm.

When the Aryans gave up their nomadic pastoral life and settled down as farmers, occupation became more specialised. Society cleaved into two broad divisions of tillers and traders. The priests and warriors, however, were top of the heap as distinct and privileged groups, on account of the prayers and protection that the others required of them.

Over time this system became increasingly rigid. One important factor in this formalisation was the Aryan interaction with the Dasas. The Aryans wanted to maintain their 'racial purity' over the dark-skinned Dasas and tried to exclude them as much as possible from all aspects of social life: intermingling, marriage, education and occupation. The Rig-Veda Veda divides society into four separate but interdependent castes. According to the Puranas, the Brahmins were born of the mouth of Brahma. They therefore mediated with the gods on behalf of mankind and addressed prayers to the gods.

The Kshatriyas were born of Brahma's arms, and were duty-bound to protect society and wield weapons. The Vaishyas were born of his thighs and attended to trade, commercial activities, and agriculture. The Shudras were born of Brahma's feet and their only purpose was to serve the other three castes. The three upper castes became known as 'svarna jati' or castes of 'good colour' or 'golden colour', an indication of Varna Aryan privilege. And caste itself devolved into a racial differentiation between the conquerors and the conquered, with colour as the most visible distinction.

Apart from the four named castes, there was another category of people called Chandalas. These were the outcastes or untouchables that were considered beyond the pale of the system. They did not live within the village precincts, nor were they allowed to enter the village, except to scavenge and collect night soil.

Members of the first three castes were called dvija or "twice-born". The second birth refers to their initiation into their caste with the thread ceremony Upanayanam The Hindus believe that the caste one is born into is based on the karma and dharma of one's previous birth Moksha. A good karma and dharma will ensure that one is born into a higher caste in one's next life. As the caste system became more rigid, the lower castes came to be oppressed by the Brahmins mediated with the godson behalf of mankind by performing sacrifices and praying.

Varna the higher castes, who increased their rituals and customs to widen the gap. The system degenerated into a harsh social order where the majority labored for the benefit of a select few. When the Constitution of India was adopted on January 26, 1950, it abolished the caste system and guaranteed equality to all citizens.

The Fundamental Rights enshrined in the Constitution ensure equality before law and equal protection by the laws to all persons in the country; they guarantee that the state shall not discriminate against any citizen only on the basis of race or caste and that no citizen shall be restricted in access to places dedicated to the general use of the public; they further ensure that the state shall not discriminate against any citizen with regard to employment or appointment to any state office and that no citizen shall be ineligible for any state office on the grounds only of race or caste. The Constitution also abolished untouchability. Its practice in any form is forbidden and Varnaconsidered a legal offence. The Protection of Civil Rights Act, 1955, further ensures this. And with the spread of education, the centuries-old caste system is inevitably crumbling.

Shudra

The lowest of the four Hindu varnas, believed to have been born of Brahma's feet. 'Shudra' is a term probably derived from the Sanskrit word 'such', to be afflicted. Many theories have been propounded over the origin of this class. When the Aryans came to India, they were already divided into three fluid classes by way of profession. As the fair-skinned conquerors, they considered themselves superior to the dark-skinned native people, who became a subject race in their own land.

The Dasas could not be left out completely; indeed, the Aryans needed them in their social structure as labourers and artisans. Accordingly, they were co-opted as the lowest caste, the Shudras, and the caste system was made even more rigid, to prevent inter-caste mobility and to keep the Shudras at the bottom of the social structure. According to the Manusmriti, these Dasas or Shudras did not have any special ability and were only capable of serving as slaves to the upper three classes. This class also included the outcaste (patita) Aryans.

A Shudra enjoyed no rights or privileges. He was not permitted to perform any sacrifices or homa, read or learn the Vedas or recite the mantras. A Shudra could marry only another Shudra. He was not allowed to enter temples and could only serve the upper three castes as a slave, barber, blacksmith or cobbler. Because of these strict and harsh social regulations, the Shudras were very supportive of the anti-Brahminical sects that evolved around the 6th century BC, like Buddhism and Jainism. Later, they also converted in large numbers to Islam and Christianity, finding solace in the egalitarianism preached by these faiths.

On January 26 1950, when the Constitution of India was adopted, the caste system was abolished and equality guaranteed to all citizens. This process of upliftment was begun in Shudra modern India by Mahatma Gandhi, who drew upon the humanitarian traditions of ancient and medieval sages, and the human rights activism of 19th century reformers, educationalists and liberal Indian rulers.

Gandhi coined the term 'Harijan' (people of God) for untouchables, who were looked down upon even by the Shudras as outcastes. He succeeded in converting vast numbers of upper class families to his egalitarian beliefs. Appropriately, the founding father of the Constitution was a Harijan (or Dalit, 'oppressed' as they now call themselves) lawyer, Dr. Bhim Rao Ambedkar. The Government of India has ever since, consciously attempted to draw these underprivileged people into the national mainstream through job reservation and special development programmes.

Vaishya

Third in the caste system (see Varna), Vaishyas supposedly evolved from Brahma's thighs. The Vaishya's duty was to ensure the community's prosperity through agriculture, cattle rearing and trade. Later, the Shudras took over agriculture and cattle rearing while the Vaishyas became traders and merchants.

From the end of the 4th century BC, as the country became politically stable, trade routes to previously uncharted areas developed. The merchant community was the first to benefit. Artisans formed guilds and co-operatives in the urban areas and guild leaders became important figures in society. Guilds also provided technical education, though formal education remained the monopoly of the Brahmins. As their economic power increased, they were expected to give alms to Brahmins, throw feasts for them, and donate generously towards the building of temples and shrines.

However, though they were "twice-born" (see Upanayanam) and economically strong because they controlled commerce, Vaishyas were denied a high social status, for which they resented the upper castes. One expression of this resentment was their support of the anti-Brahminical sects that developed around the 6th century BC, like Buddhism and Jainism. Later, however, many influential Vaishyas were honoured with titles like 'Nagar Seth' (chief merchant of the town) and 'Tazimi Seth' (privileged to wear an anklet in the king's presence and in royal durbars) Members of trading communities originate mostly from the West Coast and from Sindh (in present-day Pakistan). They are commonly known as banias, a distortion of the Sanskrit word vanik meaning "trader".

In southern India, the Chettiars and Mudaliars are prominent Vaishya communities, who have contributed to society by building hospitals, endowing universities and developing industries. While the fortunes of the other castes have fluctuated with politics and invasions, the Vaishyas alone have preserved their social and financial stability down the ages. Many of them have funded wars and political movements, notably Seth Amirchand Daga of Bikaner (Rajasthan), who is believed to have financed the British in the Crimean War. In the last century, some of the most prominent Vaishya merchants and philanthropists came from the region of Marwar (Jodhpur), Gujarat and Sindh. Many of them funded Mahatma Gandhi's freedom movement and then evolved into modern India's leading industrialists.

Kshatriya

Kshatriya literally means "protector of gentle people". Second in the social hierarchy of the caste system (see Varna), the Kshatriyas were kings and warriors. They were said to have evolved from the arms of Brahma, signifying that their role in society was the protection of people and livestock. The Hindus maintain that only a Kshatriya had the right to rule, though Brahmin rulers are not unknown. They were supposed to be brave and fearless, and to live and die by a code of honour and loyalty. They could eat meat and drink liquor and their most exalted death was to die in battle.

The Puranas say that Parashurama in his time destroyed all the men of the Kshatriya race. Since there was no one left to rule, anarchy prevailed. The Brahmins realised that a class of rulers was required. They held a sacrifice near Mt. Abu (Rajasthan), and from the fire pit emerged four clans: the Parmars, Chauhans, Parihars and Solankis. These fire-born clans Agnikula) were called Rajputs. Therefore, according to the Puranas, other than these Rajputs, those presently known as Kshatriyas are progeny of Brahmins who had married Kshatriya women. There are also leading Rajputs clans that claim descent from the sun and the moon and accordingly labelled themselves as Suryavanshi and Chandravanshi. One theory holds that in the 6th century AD, after the death of arshavardhana, the powerful ruler of Thaneshvara, there was a power vacuum in north India. To maintain the social order, the Brahmins legitimized clans of invading adventurers from Central Asia as Agnikula, Suryavanshi and Chandravanshi Kshatriyas.

Kshatriyas dominated Indian political life down the centuries. Even under the British, there were approximately 600 Indian states. These states had not been annexed by the

British and were responsible for their internal administration. Some accepted suzerainty of the Crown and allowed the British to administer their defense and external relations. When the India Independence Act, 1947 was passed, it decreed the lapse of suzerainty of the British Crown.

The Indian states regained the position they had held prior to the assumption of suzerainty by the Crown. However, in their own interest, most states merged with either the Dominion of India or Pakistan. All 552 states located within the geographical boundary of the Dominion of India acceded to it by August 15, 1947, except Hyderabad, Kashmir, and seven other states. Subsequently, these states also joined the Indian Union. Almost all the rulers voluntarily ceded all powers of governance to the Dominion government, retaining only certain personal rights and privileges.

Special provisions were made only for Kashmir. All rulers, including those of Jammu & Kashmir and Hyderabad, adopted the Constitution of India by special proclamations.

With the Constitution 7th Amendment Act, 1956, the special provisions relating to these former princely states were also omitted. The Indian states lost their separate identity and became a uniform part of India. In 1971, the Privy Purses guaranteed by the Government of India were abolished by Parliament and modern India bid a formal farewell to the princely order.

Many scions of former ruling houses are still well regarded by the people of the area. Many have joined the modern political system and contest elections to the state assemblies and Parliament.

Brahmin

The caste of priests, of those who know and repeat the Vedas. This is the first of the four Hindu castes (see Varna) and belongs to the dvija or twice-born (see Upanayanam) category.

According to the Puranas, Brahmins were created from the mouth of Brahma so that they might instruct mankind. This was why they were considered the highest of the four castes, as they had the most to do with intellect. Since it was recognised that knowledge is the only thing that remains with a person throughout life, Brahmins, as teachers, were duly respected.

In ancient times, this name was specifically applied to those priests of the Atharva Vedas (see Veda) whose place was at the northern side of the altar during sacrificial ceremonies.

The Brahmins conducted the daily rites, the purification ceremonies (see Sanskara), sacrifices and taught the Vedas. Their life was divided into four stages (see Ashram) Brahmins had to maintain a strict code of conduct and exemplify ideal behaviour.

Which they had to go through in the prescribed order. Since they were the teachers, preachers and priests they had to be proficient in sacred knowledge through the Vedas. They had to maintain a strict code of conduct and exemplify ideal behaviour. They were to be kind and gentle. For this, they earned certain privileges. They were regarded as the highest class and respected by all, to the extent that they were treated almost like gods by commoners and kings alike. They were given special gifts, dana, dakshina and were free from oppression and immune from capital punishment. It was considered to be the gravest of sins to kill a Brahmin. Only a priest could partake of the sacrificial Soma and eat the remains of the sacrifices, for no one else was regarded holy enough to eat the divine leftovers.

Initially, a person became a Brahmin on the basis of his knowledge of the Vedas. In time, the Brahmins began interpreting laws to their own advantage to maintain their privileges. And because of this unassailable social and moral authority, they were unchecked.

Thus, according to the Brahmins themselves, they were the chiefs of all created beings, entitled to all honour. The Shatapatha Brahmana declares that there are two kinds of divinities: gods and Brahmins who have learnt the Vedas. Only the Brahmins were allowed to read and teach the Vedas and hence monopolised the privilege of priesthood in the later Vedic period. They also established that one was born into a caste, which could not be changed except to be outcaste. This became the common practice, which survives till the present.

Today most Brahmins are engaged in other secular activities. However, only a qualified Brahmin can conduct religious ceremonies. These professional priests are now called Pandit, Bhatt or Shastri. There are several subdivisions within the Brahmin caste. Earlier, the different Vedas were preached by Brahmins belonging to specific gotras. The subdivisions are therefore based on their gotra or on geographic location. Presently there are 1,800 subdivisions. A Brahmin should be able to trace his origin to one of the seven great sages-Kashyapa, Atri, Bharadvaja, Gautama, Vishvamitra, Jamadagni and Vasishtha. Of these seven, Vishvamitra is said to have been born a Kshatriya and later became a Brahmin after performing severe penance.

Geographically, Brahmins are divided into two main groups-of the north and of the south, each of which has five divisions. The categories of the north are Kanyakubja, Sarasvata, Gauda, Mithila and Utkala. Those of the south are Maharashtra, Telinga, Dravida, Karnataka and Malabar. Each group claims to be the highest category. By common perception, Brahmins involved in the funerary rites (see Antyeshti) are considered the lowest category.

Bhut, Pret and Pischas

Bhuta, Preta, Pishacha A common Hindu belief holds that the spirits of men and women who died with their wishes unfulfilled, wander in the world and haunt the living instead of going to Yamapuri (see Moksha). These spirits can be broadly categorised into three classes: Bhuta, preta and pishacha.

A bhuta is the spirit of a man who died a violent death either by accident, suicide, or capital punishment, and has not had a proper funeral ceremony.

A preta (literally departed, deceased, dead) is the spirit of a dead person before his funeral rites are performed. However the word is more commonly applied to the spirit of a deformed or a crippled person or of one defective in some limb or organ, or of a child that dies prematurely, owing to the omission of ceremonies during the formation of the embryo (see Sanskara). A preta is not necessarily wicked or malicious towards people.

A pishacha is a demon created by a man's vices. It is the ghost of a liar, drunkard, adulterer, criminal, or of one who has died insane. There are many tales and fables about these spirits, describing some as malevolent and others as good-natured and helpful. Spirits are believed to live either at the site of their death or in secluded places. Abandoned homes and peepal trees are two favourite spots. The Hindus also believe that if a person goes too close to a spirit, or if the services of a professional are employed, these spirits can enter human bodies. The spirit could enter through any of the nine orifices of the body. A possessed person is said to fall sick, die, be unhappy, lose his wealth, or behave oddly.

In such cases an ojha (exorcist), through mystic rites, tries to 'talk' to the spirit inside and asks it to leave. If he knows the identity of the spirit, he asks its family members to perform certain ceremonies to pacify the spirit. This system is still prevalent in some of the parts of India. Diseases and other upheavals are sometimes attributed to the fact that a deceased family member's funeral rites have not been properly performed. To correct this, a tirthayatra Tirtha is undertaken, and proper shradha is performed.Sometimes, especially during ceremonies, a person is believed to become possessed by the spirit of a deceased family member who is either angry about something, or has come to take part in the festivities. A puja is performed to this spirit. Meanwhile the spirit is believed to be able to predict natural calamities, births and deaths through the possessed person. Hindus wear talismans, lockets, bangles, and other adornments that are believed to have the power to protect from the gaze of spirits. The recitation of certain mantras is believed to have a similar effect.

Nazar Utarana

The ridding of the effects of the evil eye' is a very popular practice. It is commonly believed that all kinds of illnesses, pains, epileptic fits and handicaps are caused by the 'evil eye', or because one is possessed by an evil spirit Bhuta, Preta, Pishacha. Unless this is nullified, the effects are said to stay. In such cases, no medication is believed to help the patient, therefore other 'remedies' have developed.

A person is said to possess the evil eye if whatever he or she looks upon is harmed. A person with an evil eye need not necessarily be wicked; usually the effect of the evil eye is unintentional. Such people do not have any distinguishing physical feature to set them apart from the rest. However, one or two 'incriminating' incidents from everyday life may doom a person to the detested category of those with the 'eye'. All those believed to be witches, wizards, and beggars are so castigated. If these people look upon any desirable object, it is believed to get ruined. If a person falls under an evil spell, there are many ways through which it can be broken. Waving a whole chilli over the person and throwing it in fire is another way. If the smoke smells of the chilli, the illness is not attributed to the evil eye or nazar. However, if the smoke does not smell of chillies, it is believed that the person was afflicted by the evil eye, whose spell has now been broken.

Nazar utarna of a more elaborate kind is performed by astrologers or professionals who do it with the help of secret and mystic rites. At times, a lemon with four or five chillies tied together, or a piece of stale unleavened bread (roti) are used for the purpose. With the help of mantras, the effects of the evil eye are. They are then either thrown away or left at a crossroad. Therefore, most people are very particular about avoiding these objects when they spot them lying at a crossroad, for fear of catching the eye if they step over them. At times these chillies are also hung on the front door to shield the house from the evil eye.

Good-looking children, young boys and girls, brides and grooms, are considered most susceptible to the eye. Small children are generally made to wear special, protective charms and lockets. Eyeliner (see Kajal) is applied to their eyes and a small black dot (kala tika) to their foreheads. This is believed to mar their beauty and make them unappealing to the evil eye.

Charms like bits of pottery from a burial ground, the dried foot of a tortoise, the tooth of a crocodile, a bristle from the tail of an elephant, a tiger's claw, or a talisman with magic mantras inscribed on it are all popular.

Some people even give ugly names Namakarana to their children. Nazar Utarnaas a pre-emptive measure against nazar. When a north Indian bridegroom leaves for his bride's house, his face is always covered with a screen of flowers, as a camouflage against the evil eye. When he arrives at the bride's house, the mother of the bride performs a ritual for the groom Arati to nullify the effects of any nazar acquired on the way. So too, a bride's mother-in-law performs the same ritual for her when she first enters her in-laws' house. Nazar is also said to affect healthy domestic animals, trees in blossom, a good harvest or fine houses.

Stone slabs inscribed and engraved with letters, characters and figures are often set up at the village boundary to safeguard the inhabitants and their cattle and crops against sickness, epidemic and disease caused by nazar. To protect their homes from the eye, women often draw mystical designs on the threshold. Black mud pots with fierce faces drawn on them are also hung on the door of It is believed that if the malefic effect of the first look is neutralized, subsequent glances will have no effect.

All these devices are believed to catch the effect of the evil eye before it affects the crops, the building, or the beings they protect. It is believed that only the first look is deadly, and once its effect is neutralized, subsequent glances will have no effect. Dhrishtamani (eye beads) are used as an indicator of the evil eye. These beads are strung together and worn by children. It is believed that if the child falls under an evil spell, the necklace breaks or the beads change colour. Rudraksha beads are also used as charms, either strung into a necklace or tied on a thread and worn on the body.

Kaal

Time. It also signifies the eternal supreme spirit Brahman. It is closely connected with the concept of salvation (see Moksha) as fate (kismat). It is also a name for Yama, judge of men and god of the netherworld.

Kaal as time is immensely important as all ceremonies and occasions are performed at a time carefully calculated to be 'auspicious'.

It is believed that if, while performing a rite, a man does not mention the lunar day, month, fortnight and the occasion, he will not reap the benefit of the rite.

Time is said to have existed even before creation. It is believed to be the source of primordial water and the progenitor of the creator. It has no beginning and no end and is like an ever-rolling wheel. There are various mythical divisions of time, such as yugas, mahayugas, manvantaras, kalpas and mahakalpas. According to the theory of the yugas, one human year (360 days according to the lunar calendar) is equal to one day in the life of the gods. One year of the gods is equal to one day in the life of Brahma, called a kalpa. A hundred years of Brahma is the life span of one Brahma, after which he dies and a new Brahma is created by the Supreme Being.

Time has also been divided on the basis of mathematics and astronomy. According to one theory, a solar day or vara is divided into 60 ghatikas (of 24 minutes each), a ghatika is divided into 60 pals, a pal into 60 vipalas and a vipala into 60 pratipalas. A pratipala is the smallest division of time and is equivalent to about 0.006 seconds. Both Shiva and Vishnu are regarded as forms of cosmic time.

The term Kaal is also used for fate (kismat). Fatalism is a popular doctrine which states that actions performed in a former life predefine events in the next (see also Moksha).

Destruction is also associated with Kaal. A particularly grim saying goes: "Vinasha kale viparita buddhi"-When it is time for someone's destruction, his fate makes him behave irrationally, with disastrous consequences.

Yoga

Application or concentration of thoughts abstract meditation, the act of joining or harnessing.

Yoga is defined as that state of mind in which the individual soul (atman) unites with Brahman, the cosmic soul. The purpose of yoga is the joining of mind and body in order to achieve perfect unity, leading to moksha, the final concentration of the spirit and its 'escape' from the cosmos.

The system of yoga, as now known, is attributed to Patanjali whose Yogasutra (5th century AD) is still the final authority on the subject even though later commentaries by Veda Vyasa and many others have been written on it.

Today, the yoga system of training is intellectual as well as moral. Before beginning training, these 10 virtues are to be cultivated within oneself: 1. Non-violence (ahinsa); 2. Speaking the truth (satya); 3. Not stealing (assteya); 4. celibacy (brahmacharya); 5. Disowning Pranayam. Yoga possessions (aparigraha); 6. purity (sucha); 7. contentment (santosha); 8. fortitude (tapas); 9. study (svadhya); 10. Devotion to god (Ishwar-pranidhan).

After practicing self-control as defined above, the student is ready for the first of the two stages of yogic training. The preparatory stage consists of asana (posture), pranayama (breath control) and pratyahara (withdrawal of senses from their object). The second stage comprises of dharana (establishment), dhyana (concentration) and samadhi (profound meditation).

Asana, or sitting down in a particular pose (there are 84 poses enumerated), is the beginning of yogic studies. The five most important postures are Padmasana, Bhadrasana, Vajrasana, Virasana and Svastikasana. These poses, most of that need training to achieve, are basis on which a person can practice meditation. Apart from their spiritual value these postures help in treating or curing many ailments. Pranayama, or the act of stretching the breath and bringing it to its original rhythm, consists of puraka (filling of the body with air), kumbhaka (holding it) and rechaka (releasing the breath). The effect of pranayama is improved health and freedom from disease. Its purpose is peace, pacification, enlightenment and tranquillity.

Pratyahara is detachment from worldly objects. It is a state of mind in which a person is unmoved by feelings of either sorrow or joy. The next stage involves dharana, or establishing a stable state of mind by concentrating and retaining breath for long intervals; dhyana, or deep and abstract concentration; and then samadhi, which means contemplating an object so intensely that one loses all sense of self and can finally achieve the status of a true yogi. Samadhi has two sub-stages, sampragyata and asampragyata. A yogi who attains asampragyata samadhi dissolves at death into natural elements and is not reborn (see also Moksha).

Apart from Patanjali's system, yoga also adopted other means like incantation and magic to achieve its aim. The most important of these are

Hatha: A technique employed by traditional dancers, which results in extraordinary mind-and-body co-ordination.

Mantra: The muttering of Mantras in meditation.

Laya: Reintegration of the mind with the cosmic soul by the practice of breath control.

Bhakti: Reintegration through love and unswerving devotion to a chosen deity.

Jnana (gyana): Reintegration through knowledge.

Karma: Reintegration through action, or doing one's duty and fulfilling one's obligations as a human being to the best of one's ability.

Raja: When the yogi loses all sense of self and mental agitation is stilled. This is the highest form of yoga.

Training in yoga is a very elaborate process. It involves the practicing of various postures (which take the major part of the day), controlling the breath in a prescribed way, concentration on a deity (sagun) or on infinity (nirgun).

Yoga also involves a controlled diet. Various texts prescribe the ideal food for yogis: barley, gruel, milk, buttermilk, fruit, roots, and grain, oil cakes and barley water. In addition to strict dietary control, many yogis subject themselves to endurance tests. It is a very common practice among yogis to bury themselves or stay in water for days on end. There are also yogis who are known to have stood on one foot for several years. Such a strenuous system of training helps an ordinary person achieve a balanced mind and body and encourages spiritual growth. The physical and psychological benefits of yoga have made it acceptable not only to Hindus but to Jains and Buddhists. The powers that a true yogi possesses are Yoga tremendous. He can become invisible at will, can stay suspended in mid-air, can dart through space with the speed of light, can still the heartbeat for sometime and can perform many other acts that seem miraculous. All this can happen only through excellent control over the nervous system.

While true yogis aim for the liberation of the soul, others practice yoga to attain peace or simply to improve their general health. There are many yogic centres in India to teach such lay students, while advanced spiritual seekers need to find and be accepted by a proper guru. Yoga and its origin are believed to date back to pre-Aryan times.

Gotra

Gotra A term applied to a clan, a group of families, or a lineage-exogamous and patrilineal-whose members trace their descent to a common ancestor, usually a sage of ancient times.

A gotra is of immense importance to a Hindu for it shores up his identity. All Hindu ceremonies require a statement of the gotra. A devout Hindu speaks out his gotra and pravara every day in the morning. Gotra also comes of use during the performance of the rites of passage or sanskaras. People of the same gotra (sagotra) are not allowed to marry, to prevent inbreeding. At weddings, the gotra of the bride and the groom are proclaimed aloud to establish that they are not breaking this socially ordained genetic precaution. Marriages between certain gotras are also not allowed; for instance, marriages between those of the Vasishtha and Vishvamitra gotras are not permitted. This is because these two sages were opponents and their descendants are traditional foes.

In olden times, every gotra had a definite task to perform. Thus every Veda had priests of specific gotras for their narration and teaching. Certain sacrifices require priests of a specific gotra only.

There are 49 established Hindu gotras. All members of a particular gotra are believed to possess certain common characteristics by way of nature or profession. Many theories have been propounded to explain this system. According to the brahminical theory, the Brahmins are the direct descendants of seven or eight sages who are believed to be the mind-born sons of Brahma. They are Gautama, Bharadvaja, Vishvamitra, Jamadagni, Vasishtha, Kashyapa and Atri. To this list, Agastya is also sometimes added.

These eight sages are called gotrakarins from whom all the 49 gotras (especially of the Brahmins) have evolved. For instance, from Atri sprang the Atreya and Gavisthiras gotras. According to this theory, the Kshatriyas and Vaishyas do not have a gotra and are to Gotraspeak out the gotras of their Purohita during the various ceremonies. However according to some Kshatriyas and Vaishyas, they are also descendants of these sages. Because of this, many a time a Brahmin, Kshatriya and a Vaishya claim the same gotra. The members of a gotra however need not necessarily be blood relations, but could be spiritual inheritors or descendants of a guru's pupils.

Among the Kshatriyas, Rajputs claim to be the direct descendants of the Sun (Suryavanshi), the moon (Chandravanshi) or the fire (Agnikula). They are also divided into numerous clans, each bearing the name of some great ancestor. Similarly every caste and class of the Hindus is divided into many clans. In olden times the members of the brahminical gotras had certain characteristic features distinguishing them from the others: the Bhargavas had their heads shaven, the Angirasas wore five braids and so on.

Gotra closely connected with the concept of gotra is that of pravara which is the invocation of Agni by the name of the ancestral sage of whichever Brahmin consecrates the sacrificial fire. As a rule, there are not more than four or five sages in one pravara. Two gotras having a common pravara are not allowed to marry. For instance those of the Kashyapa and

Shandilya gotras cannot intermarry because they share the same sage, Asita, in their pravaras. Abhivada, the formal proclamation of one's ancestry, is only enjoined upon 'twice-born' (see also Upanayanam) males. It includes not just the gotra and pravara but, for a Brahmin, the Veda to which he 'belongs', the Sutra or Vedic interpretation favoured by his gotra and finally his own name. It is said with the thumb behind the ears, in the traditional gesture of invoking ancestors. Today's great classical musicians invoke their gurus similarly before beginning a concert.

Vasishtha, Maitra-Varuna, Kauṇdinya Trayarishiya gotre, Apastambha sutra, Yajushtagadhyayi, Sri Parashuraman Sharmana aham asmiboho. In the Rig-Veda (see Veda), the word gotra means a 'cowpen'. In the Vedic period, marriage within the family was a common occurrence. The term gotra was used in its present sense for the first time in the Brahmanas. It was systematised by about the 4th century BC to accommodate changed social rules and laws and by the time of the Sutras, it was a well-established system. Even today almost all families abide by its rules.

Panchang

Literally 'five limbs' (pancha, 'five' and anga, 'limb'). The almanac of the Hindus, so named because it deals with the five Hindu divisions of time An 'auspicious' moment is considered very important, for starting a ceremony, a journey, a new venture or the commencement of studies. The panchanga is used to find such 'auspicious' moments when the various permutations and combinations of the five units of time are found congenial. When performing a ceremony (see Sanskara), the time must harmonise with the horoscope (see Janmapatri) of the person it is being performed for. Such calculations are therefore highly individual.

The panchanga is also useful in predicting the dates of various fasts and festivals like Janmashtami, Diwali and Holi. This is so because the Hindus follow a luni-solar, not solar calendar (see Hindu Calendar). Therefore festivals do not fall on the same date every year and have to be calculated. There are many regional panchangas in use across India, to determine the dates of regional festivals. But some specially published panchangas, freely sold in markets, are used nationally for declaring government holidays. Such panchangas for general use are usually made in Bombay or Pune.

The positions of the stars in each area is taken into consideration while making these panchangas. However, if religious accuracy has to be preserved, there must be separate panchangas for every 10 to 15 miles because the position of the stars and the other relevant details change. The concept of different units of time dates back to the Veda. Prataha (morning), Sayam (evening), Madhayadina (midday) and other words for the times of the day are found in the Rig Vedas (see Veda). There have been many astronomers in India, who has tried to explain time concepts, of which Aryabhatta (5th century AD) is the foremost.

Panchanga The five limbs of the panchanga are: 1. Vara (solar day) 2. Tithi (lunar day) 3. Nakshatra (lunar asterism) 4. Yoga (conjunction of planets) 5. Karana (half of a lunar day) Vara (solar day), literally means a weekday. This is the time from one sunrise to the next. For official purposes, one day is reckoned from midnight to midnight. The solar day of the Hindus is divided into four parts: divas (day), ratri (night), sandhya (morning twilight), sandhyansha (evening twilight). One solar day is made of 60 ghatikas and also of 15 muhurtas. The Hindus follow a seven-day week system and each day of the week is a vara.

The seven varas of a week are Ravivara (Sunday), Somvara (Monday), Mangalavara (Tuesday), Budhvara (Wednesday), Brihaspativara (Thursday), Shukravara (Friday), and Shanivara (Saturday). Each vara is special to a particular presiding deity. The seven varas are also associated with the seven planets of the Hindus. Ravivara is sacred to Surya hence sun worship is enjoined on this day. Somvara is sacred to the moon. As the moon is an adornment of Shiva, it is also sacred to him. Since he is an ascetic, people observe fasts on this day to please him. By performing a vrata for 16 Mondays, it is believed that all wishes are fulfilled.

Mangalavara is sacred to Mars. It is named Mangala (auspicious) to counter its malefic nature. It is sacred to Hanuman, the monkey god who helped Rama recover his wife Sita from Ravana (see Ramayana). Hanuman is said to have also freed the nine planets from Ravana's hold. Mars and Saturn are the strongest and most malefic of the nine. Devotees therefore believe that by praying to Hanuman on Tuesday, the inauspicious effects of Mars are overcome, for, by freeing Mars, he proved that he was the stronger of the two.

Panchanga Budhvara is sacred to Mercury, the lovechild of the Moon, and Brihaspativara or Guruvara to Jupiter, the preceptor of the gods. Shukravara, in turn, is special to Venus, the puissant and wise guru of the asuras, while Shanivara is the dreaded day that belongs to Shani (Saturn), whose baleful glare causes untold harm (see also Janmapatri). Tithi (lunar day) is defined as the time taken by the moon to gain 12 degrees on the sun. The moon takes about 30 days (one lunar month) to go around the earth's ecliptic. In each tithi, the moon travels 12 degrees ahead of the sun (*i.e.* if the sun and the moon are present in a specific position relating to the earth, after one tithi, the moon would be ahead of the sun by 12 degrees) hence completing 360 (12 degrees x 30 days) degrees in a terrestrial month.

In one month, there are 28 tithis, one poornima or full moon and one amavasya or new moon. The first tithi begins after the amavasya. There are 14 tithis in the shukla paksha (light half) and 14 in the Krishna paksha (dark half) of a month (see also Hindu Calendar). The names of the 14 tithis are Prathma (first), Dvitiya (second), Tritiya (third), Chaturthi (fourth), Panchami (fifth), Shashti (sixth), Saptami (seventh), Ashtami (eight), Navami (ninth), Dashmi (tenth), Ekadashi (eleventh), Dvadashi (twelfth), Triodashi (thirteenth), and Chaturdashi (fourteenth). Because the movement of the moon is irregular, a tithi ranges from 54 to 65 ghatikas. Hence when a tithi begins at sunrise and stretches to 60 ghatikas, it is equal to a solar day. But at times there are two or sometimes three tithis in one day and conversely one tithi might extend to three days. The former case is considered to be auspicious while the latter is not good for occasions like marriages or marching on an invasion. This concept of an auspicious tithi for all occasions is about 3000 years old, while the word tithi itself first came into use around 300 BC.

A full moon night of any month is considered especially auspicious. Various fasts and ceremonies are hence performed on this day. Amavasya on the other hand is not particularly beneficial, though both the new moon and the full moon are favoured motifs in classical poetry and the lives of saints. While Jyeshtha poornima is celebrated as Buddha poornima, the birth anniversary of the Buddha, Gurparb or Guru poornima in Kartik is the birthday of Guru Nanakdeva, the founder of Sikhism. And in Krishna lore, the Bhagavata Purana has it that it was on the night of Sharad poornima (the autumn moon), that Krishna invented the ras-lila, the mystic dance in which he magically partnered all the gopis (milkmaids) of his adopted clan at once, which is still an inspiring subject in dance, music and painting, and is ritually re-enacted on Janmashtami.

Nakshatra (lunar asterism) is a cluster of stars lying in the path of the moon. Just as the ecliptic is divided into 12 solar mansions (zodiacs), so is it divided into 27 'lunar mansions or asterisms', called nakshatras. The moon travels through all of these clusters in about a month. It takes a little more than a lunar day to travel into each nakshatra and so it becomes full in a different nakshatra every month.

The sun travels in about two and a quarter nakshatras every month. The nakshatras have been named according to which nakshatra the moon was full in, in the month and year the present luni-solar calendar was instituted. The first 14 are called Devanakshatra and the next 14 are Yamanakshatra. They too have presiding deities, and the Vedas say that these nakshatras are the heavenly abodes of pious people and great devotees. Yoga is defined as the time taken by the sun and moon together to accomplish 13 degrees 20 minutes of space. It is calculated from the sum of the longitudes of the sun and the moon and has no astronomical backing.

The first yoga occurs at 13 degrees 20 minutes, the second at 26 degrees 40 minutes and so on. There are 27 yogas in all, in a month (making a total of 360 degrees). These are called the nityayoga (ever occurring) and yield good or bad results according to their names. They too have deities presiding over them. Besides the 27 nitya yogas, there are special conjunctions of week days (vara) with certain tithis, nakshatras or planets, and rashis (see Janmapatri), which are also termed yoga and are auspicious or inauspicious according to various factors.

It is important to note here that this unit of time has no connection with yoga, the system by which the mind and body unite to integrate with the cosmic soul. Karana is half of a lunar day or tithi. Therefore there are two karanas in one tithi and 60 in a lunar month. There are 11 names for karanas, of which seven are termed as moving or chara and four are stationary or sthira. The seven moving karanas occur eight times a month here as the four stationary ones occur only once a month. Both karanas of a tithi are not shown in a panchanga, only that one whose end does not coincide with the end of the tithi. Like the other limbs, the karanas are also presided over by the various deities and are good or bad accordingly.

Janmapatri

Nativity chart; horoscope scroll, which records the year, month, date, tithi (see Panchanga), configuration and relative position of the planets, stars and constellations at the time of birth. It also contains the name of the person and acts as a table of fortune throughout life. Most people believe that the destiny of a person is affected by the position of the stars and other heavenly bodies at the moment of birth. Therefore extreme care is taken to note the exact time of birth to make a correct janmapatri. This janmapatri is made by the priest or jyotishi whose trade it is. He makes it by consulting the almanac (see Panchanga) and making mathematical and astronomical calculations. It is first read out during the naming ceremony or namakarana of the child. It tells of favourable and unfavourable periods in a person's life, his profession, character, marriage and progeny.

Many factors are considered to determine the fate of a person. The janmapatri gives the position of various planets and the zodiac (rashis) at the time of birth. By comparing the astral positions at birth with those of the present and future, the compatibility of the two can be judged and the future predicted. The various combinations of the zodiac, planets and time units (see Panchanga) determine the good or bad result of an event, should it then occur. The janmapatri is used in all spheres of a person's life. With its help, the yearly horoscope (varshaphala) is made which tells of the year ahead. Its most important use, however, is before a marriage when the horoscopes of the prospective bride and groom are matched. Astral incompatibility precludes marriage. A janmapatri is depicted as either a circle or square with 12 'houses' or "ghara", numbered from one to twelve. Each house represents one sign of the zodiac.

The earth revolving on its axis goes around all the 12 signs of the zodiac in one day. It takes about two hours to cross from one sign to the next. The sign rising on the horizon at the time of one's birth is the "lagna", which is also the first house of the janmapatri. The subsequent houses are occupied by the signs of zodiac that follow. Thus, if the "lagna" of a person is Cancer, it would occupy the first house. The second house would be occupied by Leo and so on. It is the "lagna rashi" that determines the character and qualities of a person. Each zodiac sign is classified as movable (chara), stationary (sthira), or of both characters (dvisvabhava). They are also classified as mild (saumya), fierce (krura), powerful during the day (dinabala), and powerful at night (nishabala). If a person's "lagna" is Leo, he is said to be of fixed character and so on.

The nine planets or navagraha, in combination with the zodiac, affect a person's life and are duly invoked in every ceremony. These nine planets are Surya (Sun), Soma (Moon), Mangala (Mars), Buddha (Mercury), Shukra (Venus), Shani (Saturn), Vrihaspati or Guru (Jupiter), and Rahu and Ketu. Of these Rahu and Ketu are 'shadow' planets which cannot be seen. At the time of one's birth, every planet except Rahu and Ketu is in one of the houses.

If a planet is present on the "lagna", it is called the major planet and affects a person the most. Planets are classified as malefic and beneficial, and each one affects the person accordingly. For instance, it is believed that a person born under the influence of the sun will possess an anxious mind, and be subject to disease because the sun is considered malefic. Anyone born under the influence of a full moon would possess elephants and horses, and be powerful, for the moon is beneficial. A person born under Mars, who is malefic, would be wounded and oppressed with fear of robbers and fire. Similar beliefs exist about all the planets.

Every planet has a 'favourite' house and a few which are not congenial. The planets with their houses are: Sun-Leo; Moon-Cancer; Mars-Aries and Scorpio; Mercury-Gemini and Virgo; Jupiter-Sagittarius and Pisces; Venus-Taurus and Libra; Saturn-Capricorn and Aquarius. Rahu and Ketu do not have a house. Of the planets, some are more powerful than the others. Natural powers in ascending order are Saturn, Mars, Mercury, Jupiter, and Venus, Moon and Sun. A planet in its own house affects a person positively. However, its presence in a noncongenial house has a negative effect. If more than one planet is present in a house, the effect depends on the congeniality between the planets and that house. Apart from this, every planet has its period of influence. The influence of the Sun lasts for eight years bestowing both sorrow and excitement. The moon's influence, which lasts 15 years, is filled with happiness. Everyone undergoes a seven-and-a-half year Shani influence, when life is difficult.

Astral conjunctions at birth affect a person's fortune inexorably. For example, if a person is born with Mars in the seventh house, it is considered the gravest of misfortunes. Such a person is termed a 'mangali' meaning his Mars planet is very strong. Not only is the person himself affected, but also so is anyone associated with him. Therefore he can only marry another 'mangali', which nullifies the effect. The janmapatri is considered extremely important even today. Many propitiatory pujas are performed to a malefic planet if a person's janmapatri indicates a bad time ahead, or to overcome astral incompatibility in a proposed marriage (vivaha).

Bibliography

Alladi Mahadeva Sastri: *The Bhagavad Gita: with the Commentary of Sri Sankaracharya*, Samata Books, Madras, 1977.

Aurobindo, Sri: *Vyasa and Valmiki*, Acharya Press, Pondicherry, 1956.

Baba, Pagal: *Temple of the Phallic King: The Mind of India*, Simon and Schuster, New York, 1973.

Basham, A.L.: *A Cultural History of India*, Clarendon Press, New York, 1975.

Basu, B. D. : *The Sacred Books of the Hindus : A Catechism of Hindu Dharma,* Sudhindra Natha Vasu, Allahabad, 1919.

Basu, B. D. : *The Sacred Books of the Hindus: Devata,* The Panini Office, Allahabad, 1917.

Bhattacharya, Haridas : *The Cultural Heritage of India,* Ramakrishna Mission, Calcutta, 1937.

Bruhn, Klaus: *The Jina-lmages of Deogarh*, MacMillan, Leiden, 1969.

Chandra Kumar, T. : *Iconography of South India*: *The Vijaya Nagar Nâyaka Period,* Tamil University, Tanjavur, 1989.

Chapekar, L. N. : *Thakurs of the Sahyadri.* London: Oxford University Press, 1960.

Clarke, Marcus: *His Natural Life*, Oxford Press, New York, 1976.

Crawford, Cromwell S.: *Ram Mohan Roy: His Era and Ethics*, Acharya Press, New Delhi, 1984.

Danielou, A. : *Hindu Polytheism,* Pantheon Books, New York, 1964.

Datta, V.N.: *Sati: Widow Burning in India*, Manohar, New Delhi, 1990.

Eliade, Mircea: *Rites and Symbols of Initiation*, Harper Torchbook, New York, 1965.

Fenton, John Y.: *Transplanting Religious Traditions: Asian Indians in America,* Praeger, New York, 1988.

Flood, Gavin: *An Introduction to Hinduism*, Cambridge University Press, Cambridge, 1996.

G. Bailey : *The Mythology of Brahma,* Oxford University Press, Delhi, 1938.

Grant, G. P.: *Philosophy in the Mass Age*, Copp Clark, Toronto, 1959.

Grewal, R. Singh: *Kundalini: The Mother of the Universe*, Singh, California, 1930.

Gupte, R.S. : *Iconography of Hindus, Buddhists and Jains,* D.B. Taraporevala Sons & Co., Bombay, 1972.

Harsh Narain: *Myths of Composite Culture and Equality of Religions*, Voice of India, Delhi 1991.

Hatengdi, M. U.: *Nityananda*, Rudra Press, Cambridge, 1984.

Hinnells, J. R. and Sharpe, E. J. : *Hinduism,* Oriel Press, Newcastle, 1927.

Ishwar Sharan: *The Myth of St. Thomas and the Mylapore Shiva Temple*, Voice of India, New Delhi, 1995.

Iyengar, P.T.S. *: Kanta Puranam of Kacciyappa Sivacaryar*, Madras, 1907.

Jacobs, Hans: *Western Psychotherapy and Hindu Sadhana*, George Allen and Unwin, London, 1961.

Jain, Girilal: *The Hindu Phenomenon*, UBSPD, Delhi 1994.

Junankar, N. S.: *Gautama: the Nyaya Philosophy*, Rachna Sagar, Delhi, 1978.

Kinsley, David R.: *Hinduism: A Cultural Perspective*, Prentice-Hall, New Jersey, 1982.

Klostermaier, Klaus K.: *A Survey of Hinduism,* State University of New York Press, Albany, 1989.

Larson, Gerald James: *Classical Samkhya*, MacMillan, Delhi, 1969.

Lewis, Oscar : *Village Life in Northern India,* University of Illinois Press, Urbana, 1958.

Martin, E. O. : *The Gods of India; Their History, Character & Worship,* Indological Book House, Delhi, 1972.

Mitchell, Robert Allen: *The Buddha*, MacMillan, New York, 1989.

Narasimhan, C.V.: *The Mahabharata: An English Version Based on Selected Verses*, Columbia University Press, New York, 1965.

Narla, V.R.: *The Truth about the Gita*, Narla Institute, Hyderabad, 1988.

Radha, Swami Sivananda: *Kundalini Yoga for the West*, Shambala Publications, Boulder, 1994.

Richards, Glyn: *A Source of Modern Hinduism*, Curzon Press, London, 1985.

Richards, John F.: *The Mughal Empire*, Cambridge University Press, Cambridge,1993.

Samdup, Lama Kazi. Dawa: *Chakrasambhara Tantra*, Luzac and Co., London, 1910.

Sebastian, J. : *God as Feminine; A Dialogue,* Peter Lang, Frankfurt, 1995.

Sharma, L. P.: *The Mughul Empire*, MacMillan, New Delhi, 1988.

Sharma, Pushpendra K.: *Hindu Religion and Ethics,* Asian Publication Services, New Delhi, 1997.

Shulman, David Dean : *Tamil Temple Myths,* Princeton University Press, Princoton, 1980.

Thomas, Edward J.: *The Life of Buddha*, MacMillan, London, 1949.

Thomas, P. : *Hindu Religion , Customs and Manners,* D. B. Taraporevala Sons & Co. Private Ltd., Bombay, 1956.

Tomory, Edith : *Introduction to the History of Fine Arts in India and West,* Orient Longmans, Bombay, 1985.

Trevor Leggett: *Realization of the Supreme Self: the Bhagavad Gita Yogas*, Kegan Paul International, London, 1995.

Tripathi, L.K.: *Position and Status of Hindu Women in Ancient India*, BHU, Varanasi, 1988.

Utpat, V.N.: *Riddles of Buddha and Ambedkar*, Itihas Patrika Prakashan, Thane 1988.

Vidhushekhara Bhattacharya: *The Agamasastra of Gaudapada*, Motilal Banarsidass, Delhi, 1989.

Vivekananda, Swami: *Raja Yoga*, Ramakrishna-Vivekananda Center, New York, 1973.

Wheeler, Robert Eric Mortimer: *Civilization of the Indus Valley and Beyond*, McGraw-Hill, New York, 1966.

Whitehead, H. : *The Village Gods of South India*, Sumit Publications, Delhi, 1976.

Yogananda, Paramahansa, *Autobiography of a Yogi*, Self-Realization Fellowship, Los Angeles, 1946.

Zimmer, Heinrich: *Myths and Symbols in Indian Art and Civilization*, Joseph Campbell, Delhi, 1946.

Zvelebil, Kamil Veith : *The Smile of Murukan*, Leiden, 1973.

Index

A

B

H

I

J

K

L

M

N

O

P

R

S

W

Y

❑❑❑